EXCELLENCE IN IT

EXCELLENCE IN IT
Achieving Success in an Information Technology Career

WARREN C. ZABLOUDIL

Universal-Publishers
Boca Raton

Excellence in IT:
Achieving Success in an Information Technology Career

Universal-Publishers
Boca Raton, Florida • USA
2014

ISBN-10: 1-62734-025-4
ISBN-13: 978-1-62734-025-0

www.universal-publishers.com

Cover image @Cutcaster.com/grafikeray

TABLE OF CONTENTS

CHAPTER ONE

Necessary Assets

No one is born naturally good at technology. Natural-born techs are a myth. In fact, Information Technology, or IT, has nothing to do with human nature. Yes, it's a product of humanity, and yes, IT is part of our modern culture, but machine based computation was never part of the human condition. It's simply a tool of convenience recently added to man's repertoire and nothing more. There's no throwing, hunting, or tackling involved in IT work…at least physically speaking. However, some nonetheless see tech work as a mental exercise not all that different from chipping a piece of flint into a spearhead. They argue that since it also involves problem solving, Information Technology taps into basic human nature and thus a person can be a natural at it.

The truth is that so much of person's mind is developed after birth that it's impossible to say how many problem solving skills were nature given and how many are developed through life lessons. What's more, the sensory skills needed to make a good flint spearhead are largely removed from the purely mental act of working with computer systems. You can hear weakness in a piece of flint by tapping on it. Touch, strength, and a fairly extensive amount of physical coordination are also part of making a first rate spearhead.

Although the spearhead maker may have visualized the finished product before he began working, just as modern system developers do with solutions they're working on, the act of turning his mental image into a material possession was primarily a physical act with some measure of problem solving thrown in along the way. It's a vastly different and much more advanced act when you turn a visualized image into a virtual outcome. A virtual outcome is measured only by an abstract input and output, which creates a product that is seen but never touched…at least not directly.

In contrast to the idea of a natural-born tech, practically all complicated jobs you'll inherit while moving up the tech food chain will involve skills you learn along the way. For instance, multitasking is a skill-set you'll definitely need to develop to stay in the game if you didn't possess it beforehand. Not just run-of-the-mill multitasking either. If you're working on more than one job at a time, then it's not merely balancing several unrelated tasks at once. You'll find yourself moving from one *group* of tasks to another in a non-linear order. That means working on task two from job B, task four from job A, tasks one and three from job C, and so on, all at the same time. Experienced techs already know all about this. You must move quickly and easily between the different tasks in different jobs without losing the perspective of each task's individual role within their respective job. Sound difficult? It's just a matter of practice. Anyone who wants to excel in IT must be good at multi-tasking multiple-tasked jobs. This can only come through repetition and experience. The ability to simultaneously track different things with multiple parts is a modern, learnt skill, and not just a part of our shared human nature, so quit pretending there are natural born techs whose skills you can never match. Everybody starts at the same place. It's just matter of how hard you work at being good at what you do. The bottom line is that everybody starts at the same level when they decide follow a career in information technology.

If there's no such thing as a natural tech, there's also no such thing as the hopeless case who can't get the hang of things either. It's always just a matter of personal effort. How much effort you bring to the job everyday defines how good you'll be at your job over time. The effort to be a good tech can be made easier by adopting four simple attributes that will help you along the way. Work to maintain these attributes every day and your career path will become easier.

The Four Attributes of IT Excellence are:

- Courage
- Focus
- Clarity
- Sense of Scope

Aspiring to these four attributes will help you enjoy a long and happy career in the world of computers:

Courage

It's safe to say that every tech will have a moment of hesitation at least once in their career when asked to take on a critical job. This's especially true when the job involves a leadership role in an area that affects many people in the company. The company's productivity will be at risk as well as the tech's credibility. If anything goes wrong, there's no place to hide. Your coworkers will know who was responsible for their inconvenience. That kind of pressure can make even the best, most experienced techs pause. Nevertheless, modern computer systems are online and running successfully all around the world, so somebody must be finding the courage to build them. The truth is that good techs have been finding the strength to build complicated and important things for many years now regardless of how critical the systems are to the people around them. Once they got started, those techs then found the confidence they needed to follow the job through to the end.

To be clear with the definitions here, courage is the ability to accept a new challenge while confidence is the bearing you maintain while the challenge is underway. The two are not quite the same so don't mistake one for the other. Don't ever think that solely one or the other will be enough to get you through a big job. Courage without confidence is starting a race you won't have the fortitude to finish; confidence without courage is being well prepared to follow a path you'll never take. Courage gets you started and confidence keeps you going. Both are only useful together.

It can be overwhelming at first to be given a tough job which affects many people who you know personally and work with every day. Take heart in the fact that there are ways to minimize the weight of the task. The first thing you must do is develop good preparation skills. For example imagine a crazy guy who, on a dare, is about to jump off the two-story roof of his house into his backyard. Just before he jumps he'll probably be more nervous than another guy with a parachute on his back who's getting ready to jump off an eighty-story building at the same time. That's because of the amount of preparation each jumper undertook. The two-story guy probably won't die; he'll just break a leg or two. What he's about to do has nowhere near the stakes of the eighty-story building jumper. Nonetheless he'll be more afraid than the building jumper at the moment their respective leaps are taken...as he should be.

A life truth is that preparation is the biggest part of courage. This applies to everything you do and goes double in the IT world. Nothing affects your courage more than appropriately preparing yourself ahead of time for a difficult job. Skill-set management, planning, and sufficient testing will all make a positive difference in your attitude before you start. If you need to take some time to review the technology first before starting on new task, then take the time needed (within reason of course). If your boss dropped the job into your lap while adding that it "needs to be done yesterday," then you need to get those communication skills going to convince him that "needs to be done yesterday" and "doing it right the first time" don't go together. If that doesn't sound easy, so be it. Do it anyway. Preparation is a large part of your professional demeanor and your professionalism is one of the few constants in the constantly changing world of IT. Preparation gives you the courage to move forward when everyone else is holding back. As skill-sets come and go, as experience on obsolete systems fade into the past (trust me, the day will come when you really can tell the new guy you've forgotten more than he knows), and as employers go through corporate changes, you must cling to your professionalism like a life preserver. It's the one thing that will keep you proud and confident. It's also the one thing that will keep you satisfied over a long career in computers. When you lack courage, your professionalism is at risk. Everything you do is vulnerable to being compromised by stronger willed people around you. Never let this happen. If you feel pressured or bullied or kicked about, remember it's the systems you build and maintain that have the final say in how good a tech you are. Build them well and keep them running and only good things will follow. No one can take away up-time from you and up-time (when the systems run well) is the only true measure of how good you are.

Still, there is the occasional worst case scenario. Let's say for example there's a rushed boss who wants you to jump into a job you know you're not quite ready for. He wants you to move on the spot and you, as a professional, prefer to take a moment to get up to speed on the technology first. The boss is so disappointed in your lack of recklessness that he finds someone else to get the job done on the double quick. To make matters worse, when the new guy takes over he gets lucky and finishes the job correctly and on schedule too. As a result, the boss starts treating you like a goat and the other guy like a hero. If this happens to you, then keep one critically important thing in mind: *you were still right*. Even if the new guy succeeded brilliantly, he was wrong to begin in the first place and relied far too much on luck to finish the job.

To pick up on the previous analogy, let's say the guy who jumped off his two-story roof hit the ground and rolled to a stop uninjured. He pops right up and is the hero of the moment, receiving "oohs" and "aahs" from everybody around. Does that mean he can climb right back up and do it again? No. What it means is that he made a bad decision and got lucky. The next time he jumps, he faces the same tall odds of disaster. The same goes for any tech who rushes into a job without suitable preparation. If fact, pity the tech whose boss is so impressed with his lack of restraint that he's now expected to get jobs done 'on the quick' every time. That's no different than expecting the guy who jumped off the roof to keep doing it over and over again on a regular basis. After all, it worked out fine the first time. If you're professional enough to want the job done right instead of just quickly, the fact is you were correct in your restraint. Your boss, who was cutting corners, is only inviting trouble down the road for the whole department.

Doing things without proper preparation will always cost the company extra money in the long run. The inevitable mistakes that kind of work ethic brings will cause more downtime and delays in worse ways than taking time to properly prepare ever will. If you find yourself in this situation, remember your boss made a bad decision based on the needs of the moment and you made a good decision based on your long-term professionalism. Don't ever doubt yourself in this regard, no matter how much it may sting your self-esteem at the moment. In fact, it's even reasonable to think that if the boss had planned things better from the start the current rushed situation wouldn't have happened.

Good preparation means understanding all aspects of the job at hand. This includes end-user requirements, the technology involved, the milestones to be managed, the risks during implementation, and the abilities of those who are assisting you. The more you can lay down a clear path to follow in terms of rolling a system into production, the more courage you'll have from the beginning and the more confidence you'll maintain along the way. Good preparation leads to courage, courage invites confidence, and confidence leads to success.

Once you understand the technology enough to proceed with a reasonable amount of comfort, the next step is to measure all the variables. These can include such things as cost, workload, time, risks, workmanship, and user orientation. The combination of variables is different with every job and you must work through them all to determine their importance in each case. Confidence can be gained by understanding how much focus each variable deserves relative to its impact on that particular job. This is because understanding something in detail helps diminish the element of surprise and, as has been proven over and over again, surprise is the one thing that can never exist in the world of IT.

Being able to understand all the variables in depth for each job is something that comes through experience. It's never a quick task. You must take care to understand as much as possible about how to proceed on a job before you start to work. Never think that you'll figure things out as you go. Planning to "cross that bridge when you get to it" doesn't work in IT. Presume nothing and prepare for everything. If you still get tripped up along the way, you'll at least be able to recover rapidly and get things back on track with less effort.

The variables that can cause you to stumble during your work are measured as risk. Risk comes in so many shapes and sizes you may not even realize it's even there until too late. It could be a loss of funding caused by a previously unannounced change in corporate structure, a component stuck on backorder with a vendor, a team member overstating their skill-sets, or a large end-user group struggling to find time for proper orientation once the new system is successfully built. Risk can come at you from every angle and at any time along the way. Being burnt by unknown risks can dampen your courage for future jobs and hurt your confidence while finishing the one at hand.

It takes imagination to understand risk. The more ways you can imagine how things can go wrong, the more prepared you'll be before you start. Minimizing risk is the same as minimizing surprise. Don't let yourself be surprised by anything as you move forward. It's not good enough to try to merely anticipate risk. You must fully understand and eliminate as much risk as you possibly can before you begin. An IT project must not be an adventure into the unknown. There can be no unseen elements hiding around unforeseen corners waiting to trip you up. You must anticipate where all those corners are and what elements of risk they might conceal before you begin.

Nothing gives you more courage than knowing how something will turn out before you start. One of the nicest things about the IT world is that you get to work on things you control from the beginning. After all, those systems didn't fall from the sky. They were built either by you or someone like you. As such, you have the power of God over what goes on inside them. If you don't fully understand them, don't start working on the job until you do. If deadlines that don't allow time for understanding were irresponsibly assigned by management, then convince management more time is needed. Running head-on into a fail-state that could have been foreseen with a reasonable amount of preparation can cripple a company and break morale. No matter how big the hurry, unforeseen risk can easily end up costing more in the long run than using an appropriate, evaluative approach.

You gain more credibility from a successful job than from a failed one, no matter how heroic your efforts were. In IT, you have the opportunity to foresee the future, at least as far as any particular job is concerned, so don't waste it. Take the opportunity to understand all the elements that can go wrong before you begin so you can move forward with all the confidence in the world. You can predict all events and vanquish all worries ahead of time if you put in the effort to do so.

Bigger jobs may also have more than one deadline. Understanding how risk affects each of those deadlines (or milestones) is a critical part of performing any job well. If the job is multifaceted, then not meeting a particular milestone can be more than just a scheduling hassle. The entire proverbial assembly line can be stopped if one of those milestones is missed. That means even the possibility of a missed milestone here or there is just one more thing that needs to be accounted for in the planning stages. It's especially true if the deadlines were poorly thought out by the staff member who assigned them. The best thing about moving from milestone to milestone is that it gives you a clear line-of-sight path to the next endpoint. If you break even the most unreasonable deadline into manageable steps, you can work your way through it with less effort than if you had simply dived in and hoped for the best. Most importantly, you'll reduce the element of surprise while working to get the job done.

Risk loves things like rush jobs and short deadlines. Those are its easiest targets. Just as water always follows the path of least resistance, risk always follows the path of least preparedness. An important skill-set you should develop is the ability to understand how to manage the risk of those occasional short deadlines. If you get a short notice job and you understand the accompanying importance and the risk involved, the first course of action you must take is the preemptive act of building up the courage to request some game rules as soon as you can, even if nothing particular has been assigned my management yet. Rule number one is that is that you'll always be given a reasonable amount of time to develop a workable plan for any new job. Rule number two is that you won't be thrown into a job cold with no skill-sets for the solution you'll be working on.

Usually these game rules can be a simple request to your boss. Bosses with any IT sense will appreciate your forethought and probably have a pretty strong idea of what you're requesting based on their own experience. Confidence, as well as courage, can also be handed to you through good management too. If you have a lousy boss who doesn't get it, the best you can hope for is good luck with the jobs you're assigned as you struggle to mitigate risk more effectively. In some cases, it's not unreasonable for you to consider mustering what little confidence you have left and use it to find employment elsewhere; after all, your ability to maintain your professionalism must *always* come first throughout your entire career.

Keep in mind that even the worst bosses can't hide for long from the high employee turnover rate they cause. Turnover rates cost the company money by making it difficult for the IT department to operate efficiently. The hard truth about IT is that no department can cut more deeply into a company's bottom line than Information Technology does. IT is always considered overhead; even in companies that sell IT services. IT doesn't generate a single penny of profit for any company, its true value can only measured by how cost efficiently it can support billable minutes for the rest of the organization. All computer related inefficiencies will eventually cause the company to lose billable minutes and suffer increased overhead. No matter how politically connected the bad IT boss may be, this truth will always catch up with them.

However, what if the project assignment is a 'one-off' anomaly? Let's say a good boss got a rush request from a customer that can't be denied. Then you need to figure out how to professionally mitigate risk in a hurry. Step one is to take a brief moment to go on the record by politely reminding everyone for the umpteenth time that this's not how things should be done and to truthfully establish an understanding of the increased risk you'll be facing. After all, it's better to get the gripes out before you begin than it is to complain later on when the deadline is approaching and things are crazy enough on their own without you adding to the noise.

Whether you're rushed or have plenty of time to work with, a handy trick to identifying all those pesky risks in any job is to narrow your view to as granular a level as is reasonably possible. The more granular you have time for, the better. Do this after first gaining a solid understanding of what needs to be done and laying out the job milestones as usual. Then divide each milestone into several smaller projects with contiguous start and endpoints. The smaller, more granular view provides a clearer evaluation of risks relative to each milestone. As each risk is identified it needs to be assessed against the scope of the job as a whole. Used properly, this trick provides a more thorough approach to risk identification by narrowing the scale of each part of the job that any given risk can affect.

Narrowing the scale not only helps you better assess risk, as a side effect it also forces you to become more familiar with the job down to the granular level you've chosen. Focusing on the subdivided parts before you begin will give you a more detailed overall picture of how to proceed. This increased clarity helps identify even the most subtle of job-related details and allows you to eliminate any risk hiding in those details before the work begins.

The more you understand up front, the more courage you'll have when you start. The more courage you gain, the more prepared you'll be to move quickly to deal with issues as they arise. This is where the confidence part comes in. Even those elements of risk that managed to remain invisible during the planning stages won't be quite as overwhelming when they rise up and try to knock you off schedule. No matter what happens after you begin, the more prepared you are at the start, the more confidence you'll have in the middle of the job in the event any setback does occur.

Preparedness is the result of both planning and training on the plan. For preparation to be considered fully complete, at least one rollback option must also be in place in case something goes wrong beyond your control. A rollback plan is not only a critical step to ensure the operational continuity for end-users; it's a real confidence booster as well. A rollback plan is the ability to turn the clock back to a time when everything was still functioning properly. It's the reset button in the video game of life. Rollback is made up of the contingencies needed to keep a failed implementation from ever reaching the end-users. A good rollback plan isn't based around a single trigger either. Rollback can be total if the implementation was a complete disaster or just partial if the implementation went mostly well, but had a few bugs along the way. Every undertaking is always

15

less scary if it includes plans for how to get back to a safe place if need be.

A final note on courage is that it's almost a certainty that every busy IT professional tech will experience at least once in their career a bad spell in one form or another when anything which could possibly go wrong with their work usually does no matter how hard they try to make it right. Even the best techs out there can experience times in their career when it appears fate is against them, or at least extending some persistent doubt in their direction. This type of thing usually happens when you're going through a "snake bit" period. That is, while everything you touched before on the job seemed to turn to gold, now everything you touch in your work seems to turn to lead. You're still a great tech, but, for some unknown reason, no matter how good you always are at making judgment calls about what to do next, for no apparent reason those calls simply start going wrong despite your best effort. When you look back at your recent work all you see are mistakes…usually dumb ones too.

This happens to everyone every now and then in the complicated and constantly changing world of computers so don't take it personally. If you're doing your best and your best is good enough, then there's no explanation for being snake bit that'll make sense so you shouldn't let it hurt your confidence. Just keep doing your best and the rest will take care of itself. It's an honest truth that every tech in the world will experience a snake bit period at least once in their career; including you. When you're snake bit, even your best work can leave a dirty trail behind you. You look back at your recent jobs and all you see are problems all around. Not big, complicated mistakes either, but irritatingly simple ones you can't believe you keep making. Since this kind of thing happens to everybody at least once over a long career, you should create a "snake bite kit" ahead of time for whenever the situation finally arrives. For starters, remember that being snake bit is always temporary. It's best to just buckle-down and push on through it. Soon enough the successes will start returning and your courage and confidence will get back on track. You can, and should, bank on it.

One tool that should be in every snake bite kit is your resume. It's old advice, but as long as you know you know it's truthful, re-reading your resume helps remind you that your self-confidence is not unwarranted. You did good work in the past and you'll do good work in the future too. Revisiting your past successes will often give you the boost in self-confidence you need to move forward with your head held high. As long as you know you're good, the successes will return. When they do, count them...literally. Always keep your resume up to date if for no other reason than to remind yourself that you are a first rate tech. That kind of continuity in your confidence over time is what will help you rise to the top of your professional world and earn a reputation for excellence in your company.

Other curative steps can include reviewing your certifications or degrees and what you did to get them. Also, any past accolades from bosses or coworkers are worth review too. This's a highly personal act so design your snake bite kit as you see fit. The thing to remember is to never lose confidence in yourself, even if it others around you appear to have begun to doubt your ability. It's OK to give yourself credit even if no one else does. In fact, it is a critical part of moving over those rough spots in your career and keeping your courage level high at all times.

Focus

Nothing defines a solution better than the number of details it involves and nothing measures a tech's skills better than their ability to work with those details. Having good focus is how you come to terms with those details. Taking solutions from a variety of vendors and combining them into a single system means adapting the nuances in each of those solutions to work alongside all the other nuances around them. Being able break something down into its basic parts so you can accurately focus on its details is a critical skill for anyone who works with complex elements for a living. In reality, the old adage, *keep it simple stupid*, only applies to jobs that were simple to begin with. The beauty of a great solution will always be found in the complexity of its details.

Many techs have a problem coming to terms with the true level of detail that good preparation entails. Those techs need to learn the hard way that overlooked details will always come back to haunt them when moving to production. You should guard against this kind of education. Don't let this happen to you. Take the time to inventory all the details involved with the job and touch on *each one* before starting to work. Maintaining a high resolution focus goes a long way toward achieving excellence in the tasks you're involved in. For a tech traveling to a site on a service ticket, this could mean going through all the possible causes of the problem in their mind before they arrive. For someone starting a new project, it means

17

working through the project and accounting for all the details in a thorough project plan before moving to development.

The more you trial your solution to account for all of its details before moving to production, the better off you'll be. This's because trials are a great way to expose the hidden problems in your system. Failure is one of the most useful tools you can have when working to get things right, so it's actually a good thing to fail during a trial. The more you understand about how things can go wrong while you're still in the preparation phase, the better everything will work when you finally move to production.

There are different outcomes that people may be inclined to call positive for any IT job. There is the top grade "right the first time, every time" outcome, the bottom grade "good enough for now" outcome, and several shades in between. It should be obvious what kind of outcome you should strive for if you want to be thought of as an excellent tech. Always remember that from your perspective as an IT professional, the definition of quality doesn't come from how well the computer system runs when you're finished working on it; that's the end-user's definition of quality. From your perspective, the definition of quality comes from the number of those exquisite details you accounted for while the job was underway. The rest will take care of itself when the job is done. The more granular your focus is at the beginning, the higher quality your output will be in the end.

If you want to be a real professional, you must always be able to recognize the difference between good work and "good enough" work, even if end-users won't ever notice it. Professionalism in any line of work is gained more through the practitioner's ability to understand the smallest details of their craft more than from anything else. No amount of happy talk will ever replace that. Whether it comes from classroom education, hands on experience, or a good resource library, the ability to maintain an accurate focus on the smallest aspects of what you're working on is what will most define you as being good at what you do. Anything less will just define you as less. It's the good techs who keep track of all those fine details that will develop the reputation of being excellent.

Clarity

It can't be said enough: the best measure of how well you understand anything complicated is how clearly you can explain it to someone who knows nothing about it. This goes double in IT. Few things are more mysterious to end-users than their computers and whatever it is their computers are tied to. The gap between the knowledgeable and the novice can never be underestimated when interacting with end-users. What's more, the definition of novice can be much broader than the average tech might think.

If you look at it from the perspective of your daily activities and how they continually create something new in your environment, then anyone, including your boss and your teammates, can be considered a novice, at least with that one thing you're working on at the moment. Your work in IT affects so many people that you're more or less required to keep all involved parties abreast of your actions to some extent at all times. This can be a status report to a boss, a quick answer to a teammate, some updates to the helpdesk, or a notice to end-users about something new coming their way. It can also be a reminder given to upper management about why the expense of something new is worthwhile. While technical skills are part of your craft, you should treat communication like an art form because it is really that important.

The critical thing to remember about practicing the art of communication is that while you multi-task multiple-part tasks all day long, you must always communicate about them serially. That means to sticking to only one topic per answer. When you're giving a status report always follow a linear order and don't jump from one task to another. While all those multitasks make complete sense in your busy mind, your listener is going to become confused in no time. It's amazing how difficult it can be for some techs to communicate sensibly when they're working on a number of different multi-tasked jobs at once. Any status updates they give are basically just a brain dump of barely connected facts.

To make matters worse, the typical tech is surrounded by other multi-tasking professionals, all giving status updates, asking advice, or anything else that involves describing their ongoing activities. It's critical that nobody add more than one topic at a time in this situation. It's hard enough to keep track of your own jobs without trying to follow along with someone else's jobs being described to you in haphazard order. Clear communication starts with simplicity. The thing to *always* do when communicating is to find the simplest, yet fully accurate, method for conveying *just* the information required for the moment and nothing more. Any elaboration should only be used as a last ditch rescue attempt when clarity has failed after a try or two. Even then, it's critical to be short about it. When elaborating, for whatever reason, remember that if you weren't clear in the first place, adding a lot of marginally organized detail later on isn't going to help the listener much. Master the art of saying just enough to be clear, concise, and accurate with your information from the start. It'll serve you well in the long and make your listeners happier, too.

The same goes for what you write. Keep written communication short and to the point. It's easy to get in the habit of putting everything into a single e-mail that leaves nothing uncovered. This can be especially true after spending time with those "but why?" end-users who persistently e-mail questions about things in a level of detail you know they don't fully understand. As soon as you answer one question, they reply with another that's equally hard to quickly explain. Pretty soon you have a long thread running down the screen of back and forth "but why" questions and answers.

Sometimes techs respond to this type of end-user by preemptively including a response to every anticipated question in one big "first strike" e-mail that covers every possible detail. Their first strike e-mail usually ends up being a bunch of paragraphs that run on forever. While this's better than blowing off the overly inquisitive end-user with "JUST BECAUSE" in bold font, it can lead to the bad habit of responding to all e-mails with a first strike format. Don't let this kind of thing change the way you answer questions in general or you could find yourself putting everything into one big e-mail even when no one is asking for it.

The irony here is that when the recipient sees all your well-chosen size eleven Calibri font words running down the screen, they may not even bother reading your e-mail in the first place. Nobody likes reading long e-mails in the middle of the day. If you have ever had someone ask you a dumb question just after you explained everything to them in an detailed e-mail, don't blame them. It's not their fault if your e-mails are too long to read. It's best to keep to short answers and avoid the long novels. While you're at it, break your e-mail up into small blocks too. Put an extra line feed in every now and then. Even when a paragraph is relatively short, it can still look too long to read if it's a single block of text. Remember to never be a hassle. You're not the only one who gets tired of reading e-mails all day long. Add a little extra formatting to make your communication easier to read. Your recipients will appreciate it.

Good communication skills don't only apply when interacting with others; they're also important when communicating with yourself. This isn't meant to be trite. Being truthful with yourself makes a big difference in how good a tech you'll end up being over time. This isn't about being hard on yourself: self-honesty has nothing to do with marginalizing your own self confidence. However, being able to communicate truthfully with yourself means understanding your limitations in both knowledge and time management. All the knowledge in the world won't help much if you never have time to use it. On the other hand, having only a few skills-sets but a lot of time to figure things out may seem nice, but you won't be getting much done. The bottom line is; being truthful with yourself will help you make better decisions about the jobs you're working on.

This particularly applies to knowing when a job is truly done. This is one of the hardest lessons for techs to learn and is the biggest failing of rookie IT professionals. For example, stress can create a deep desire to just tip-toe away from a quick fix in the hope that the problem won't pop back up the moment you leave the area. Or a tough time management issue might have you running off to the next job before the current task is really finished. Sometimes less experienced techs don't know enough about what's being worked on to feel comfortable with sticking around to confirm that the work they implement will take in the long term. As every tech eventually learns, calling a job done too early will lead to repeat visits to work on the same issue down the road. The worst side effect of this is the end-users will eventually start to question your abilities.

Learn to deal with this situation by communicating with yourself well. First, listen to that frustrated voice in your head that tells you that this time you really will crack down and study what's needed to be proficient on the systems you're responsible for. When you ignore that voice, you're ignoring common sense. Get to know that voice and understand what it sounds like. If you want to claim a good level of street smarts on the job, it's your common sense that'll give them to you. Ignore common sense and you could find yourself both miserable and unreliable. Many IT professionals, if they give themselves a chance, can be surprised at how often their first notion turned out to be the correct one and how much it helped them strive into a new job with maximum effect by allowing them to move quickly from the beginning.

Striving into a job helps you to make a better call about when the job is truly done. In fact, it's the biggest part in getting things done right "the first time every time." If you want to gain excellent results with the tasks you're given, you must always listen to your common sense when it tells you a job isn't quite complete. Always keep in mind that it's you alone who brings the last measure of your work ethic with you wherever you go. Never mind your boss. No matter how much your boss might threaten you, that source of stress only accounts for about 96% of your best effort on a good day. The remaining 4% of effort, the part that makes you an excellent technician, is something you must bring to the job on your own. At some point you must move beyond the simple job order and take ownership of the work you do. It's in the pride of ownership that you find that final 4% of effort. Never settle for the 96% solution, even if it makes your life easier for the moment. Let your common sense give you the clarity you need to cover that final 4% on your own and you'll gain excellence in no time.

Scope

If you want to be a superstar, or at least the best tech in your department, taking single steps at a time is the best approach to follow…no matter how ambitious you are. If you're addicted to accolades or just like having a say in how things are done, the status of an excellent worker is how you can achieve your goal. The important thing to remember is that you must never overreach or your efforts will end up doing more harm than good. Never lose track of how much you can legitimately do and only push a small step beyond that limit each time you're ready to move ahead in your career.

Taking on a job too big for you to handle or taking on too many jobs at once can damage your credibility if things go wrong. On the other hand, stepping forward to volunteer for the tough jobs on a regular basis is a great way to move your career upward. You must always temper your ambition by taking careful measure of your current ability; otherwise you'll end up biting off more than you can chew. How you balance the need to excel with the common sense of being careful about what you're willing to get yourself into takes some experience. It's all about understanding your reach.

Never mind your physical arms; they aren't part of the reach being discussed here. Instead, try imaging you have virtual arms with the strength of a small construction crane. You can reach those arms way out and pick up very heavy things. However, like any crane operator, you still need to know the limits of that reach or you'll tip yourself over. This limit can be called your professional reach, or more accurately, your extensible reach. Your extensible reach is the combination of your courage and confidence, your ability to focus, your effort to train hard, your willingness to take the lead when called upon, and what you can actually can do today, all added together. It's the blend of these things when taken as a whole that give you the proper sense of scope when taking on new assignments.

Going beyond your extensible reach will always lead to things which are best avoided. For example, even the best techs can only multi-task so much before they realize they're spending more time moving between jobs than they are on the individual jobs themselves. It's not always a clean jump from one task to the next. Knowing your limits here is critical. If you find yourself constantly stopping to review before you can move to the next item on your list, you're doing too many things at once. It's not that occasional review is a bad thing, because it isn't. However, constant review means you're struggling so much to keep tasks organized that you're losing efficiency. If you're juggling so many multi-tasked tasks that you keep forgetting what each one is about, then it's time to slow down.

Even if it means telling the boss you're too busy, don't feel bad. Going beyond your reach can make you a liability to your department by putting you in the position to make bad choices. Trying to do too many jobs at once can cause you to replace good work with marginal work. It's far better to deal with credibility issues caused by a lack of time or knowledge than it is to work through credibility issues caused by major mistakes.

Some of the blame may also go to managers who aren't allocating resources thoughtfully; however, you shouldn't use this as an excuse. You must find ways to take charge of your workload well enough to make it through each day with balance. Ultimately, it's your career to manage and no one else's. Trying to gain accolades through reckless behavior could lead to mistakes that destroy your chances for dream jobs down the road. Mistakes have a way of following you around, even if you've hidden them from your resume. Have the strength to take charge of your career and don't let anyone, even a boss, talk you into recklessly going beyond your extensible reach.

While the courageous tech will take opportunities when given, the excellent IT professional will only take jobs he is certain he can deliver on. Know the difference between good courage and reckless courage. Overreaching and gambling are both pretty much the same thing; except that in IT you're gambling with your career as well as your department's credibility in the eyes of your company.

A good way to prevent overreaching is augment your current skill-set before taking on a wider range of tasks. After all, there's really no use in being able to do a variety of different things marginally well. Being able to do a few things really well, and then gradually adding to that list of things, is a better long term approach to a successful career. It's also an easier way to handle skill-set management. You should try to avoid being too much of a generalist, even if you're running a small network on your own.

Certifiable skill-sets add confidence and will create success. If continuous skills training seem tedious, that's because you haven't felt the pain and embarrassment that comes from totally screwing something up. Mess up an important system a couple times and all that studying won't seem quite as painful anymore. Stay ahead of the knowledge curve by anticipating as best you can what it is you're going to need to know in the coming year and start running down the learning track early. Not only does this take away much of the pressure which comes from trying to learn in catch up mode, it also puts you in a good position to be one the first to volunteer for new jobs that come along. Preemptive learning is the safest way to push the envelope and extend your extensible reach.

Always begin a job with a strong foundation. Never pretend to have skills that you don't and never plan to learn how to do something after you've already volunteered for it. Both are paths to disaster. It's one thing to drop the ball on something you're familiar with because you can always recover yourself and do it correctly on the second try, but if you're working in an area that's new to you and you don't have solid knowledge about it, then there's nothing to fall back on. There's no deep understanding of what you did wrong to help you recover quickly because you have no real understanding at all.

The best way to stay in front of the knowledge gap is to maintain the scope of your career within a particular area of expertise. When you first start a career in IT, you'll find there are the many paths to choose. You can go into network engineering, application development, WAN storage, software programming, disaster recovery, telecommunications, systems security, systems administration, and so on. Each is different from the next, but within those diverse career areas are even smaller, more specific career paths too. You can broaden out a bit as the years go by if you think it makes you more employable, but try not to broaden your scope too much. If constant learning and knowledge maintenance is tedious for just one area of IT, imagine how difficult it can be to manage constant learning across several areas of IT. This'll be covered in more detail later in this book, but suffice it to say if you aren't careful there could be a point when you'll have too many things going on with your career to reasonably keep track of.

Most IT professionals find it's far better to concentrate their skills on just one or two paths at most. Being a jack of all trades doesn't work so well in IT. Begin by finding a niche to excel in to start your career foundation and then scale out as you go. If you're new to IT, then this niche might be even handed to you by a boss. Even if this task is not your first choice and you had planned to go in a slightly different direction with your career, do a good job anyway. If the skills don't stay with you after you move on to something new, the professional work ethic you maintain will. Always have a strong sense of ownership with anything you undertake. This includes stepping forward for new tasks only when they're doable in a professional manner. Professionalism comes first, all else follows that.

Even with masterful skills you can find yourself in trouble if you try to do too much at once. It's important to know when your extensible reach is at its limits. Do this by honestly taking an assessment of how complete your 'finished' jobs really are. After all, fixing a tricky bug isn't as easy as changing a tire on a car. Finished and complete don't necessarily mean the same thing in a world where the solutions are often virtual. For the sake of your own long term sanity, you must always resist the impulse to call a job done before it really is. That means get all tasks completely finished by everyone's measure and not just your own.

Leaving incomplete jobs for others to clean up will give you a reputation as the kind of tech who can't be trusted to work on their own. Having someone follow behind you on a regular basis to check your work is no way to achieve a reputation for excellence. Don't fall into the trap of thinking you'll find time to make corrections later either. That's just wishful thinking. The fact is there'll always be more jobs later on that'll take your future time and effort to complete. If you can't get each one completed when you have the chance, you probably won't have time to come back later to finish up. If you reach a point where this's how your workload feels, it's time to slow down and take stock of your situation.

Remind yourself that your professionalism is at stake and that your work ethic should never be compromised. Not doing a 100% job because you think non-techies won't notice your short-cuts is a poor way to work. The discipline that all techs share requires a standard of excellence that can be hard to maintain but is too important to let slide. Company employees who don't know anything about computers are depending on you to lead the way. Don't short change them. Know your scope and don't go beyond your limits until you're ready to move on to the next level.

CHAPTER TWO

Bad Habits

No matter how life-defining a career may be, it often follows a less glamorous path than the one dreamed of in childhood. A person can aspire to be a doctor but leave medical school to end up in advertising. Another can learn to work on fast cars from a favorite uncle only to become a school teacher instead of a famous race car driver. Still another might join the army for a temporary stint fresh out of high school and end up retiring decades later as a gray haired officer with a great military legacy. Any path that a person stumbles on can become a life calling, even if it was never their original intention.

This's how it has always been. So it's not surprising that some people don't care as much as they should about the quality of work they do. They show up at their place of employment and sit in their cubicle, stand at their station, walk their beat, or whatever, and move through their day one minute after the next, never giving any thought to how they can improve their performance or how their performance affects others around them.

As a computer professional you can never fall into that mindset, even if you never dreamed of working with computers when you were a kid. It's an old adage that there's no such thing as a bad job, only jobs that are badly done. Certainly the IT world is full of badly done jobs. Some IT service providers seem to be getting more adept at disappointing their customers every day. People regularly complain that calls to customer service mean confusing voice menus, long hold times, inconsiderate staff, marginal repairs, and so on. Nevertheless, just because badly done jobs seem to exist all around you, it's no excuse for you to allow poor performance to become part of your career too.

Something you should keep in mind is that great buildings are never built by as many people as they eventually occupy. A few well motivated people can make a big difference in the lives of a much bigger group of people around them. It's an important truth that anyone who rises a bit higher in their daily job can help raise their co-workers' efforts a bit too. In any economy, old or new, every little bit helps. The idea is to accept the career path you're on (even if it wasn't your original choice) and stay true to it by rising to the calling it presents, even if others around you haven't. This's where the approach you take when working on computers starts to matter.

As an IT tech, you have a workplace impact that extends far beyond your desk. How well you do your job affects the lives of all the people working around you. Being a tech is not an assembly line kind of job where you only affect a small part of a whole. You can directly impact the careers of everyone else in your company by how well (or how badly) you handle your own career. Everything you design, everything you build, everything you maintain and upgrade...in fact, virtually everything you touch in your working environment...will be used by everyone else in your company to earn their living. This is not a small responsibility.

Your coworkers' ability to make their mortgage payments, their car payments, their kid's tuition payments, and even get that oh-so-rare Christmas bonus, can be immediately impacted by how well you do your job in the IT department. If you've ended up choosing to work on computers for a career, what you've really chosen is to help provide and maintain a large part of the foundation on which the world operates. Sure this may seem an overstatement at first glance, but it's the flat truth. You must never forget that how well you do your job in IT on a daily basis determines how well your coworkers will be able to do their jobs on a daily basis too.

Unfortunately, experience is beginning to show that a few sloppy computer techs might be finding employment in the ranks. It sounds cranky and cynical coming from an old tech...and yes, this's only the opinion of one person...but in a world where virtually everything is done on computers nowadays, marginal computer support really does cause many people to suffer. The world of IT should never be taken lightly and there's a no greater sign of a marginally conceived or executed solution than the struggles of the end-users using it.

People who work on computers all day long get to know them on a personal level. Every little glitch, each sticky part, and anything slow will be noticeable and remembered. The company's employees should be able to remain focused on the work they're doing and not on the tools they're doing it with. Any small operational defect will slowly build into a major source of frustration, or worse, into invisible operational overhead that costs the company in ways that management can't easily see. As a computer tech, you really do have this much impact on the life of everyone around you.

Before covering the characteristics that'll lead you on a path to achieving excellence in your job, it may help to first outline the common characteristics that lead to something less. There are many types of bad behavior in the IT world, but nine in particular seem to be the most prevalent. If you find yourself developing one of these nine characteristics in your daily work ethic, you should look in the mirror the very next morning and think hard about ways you can become better at what you do for a living.

The Nine Bad Tech Types are:

- Whiners
- Too Busies
- Experimenters
- Constant Rookies
- Divers
- Know-it-alls
- Fraidy-cats
- Nerds
- Breathless Wonders

These labels may seem a bit trite, but each is an accurate description of the person involved. In each case the tech isn't only performing in a way that's detrimental to him or herself, but also in ways that can slow their team down, create wrinkles that shouldn't exist in an otherwise good project, or even affect client relations. If you recognize any of these traits in yourself it would be a good idea to take a contemplative step back and review the approach you've taken so far in your career.

Whiners

Whining is done by techs who want others to believe that something bad isn't their fault. The lack of desire to accept responsibility for the occasional negative outcome is a common trait in all people, but the IT whiner brings this trait into the LAN room. The excuses can be routine, too. Either they were overworked, under-informed, or misguided in some way that made it impossible for them to get the job done right.

Even when it really isn't your fault, you should never whine. Taking responsibility when something isn't right (even when others know it wasn't truly your fault), is a great way to build credibility. It makes you the tech who steps up and is accountable for the jobs you take part in. This isn't to say that you should volunteer to be a sacrificial lamb whenever someone else screws up. What it means is that how well you can carry yourself when working through a tough spot defines how much everyone else can trust you in the long run. Whining that something wasn't your fault when everyone knows it was is a fast way to lose the trust of others. No matter how the problem came about, if you were the responsible party when it happened, then it was your fault. Whining will only make things worse, not better.

Moreover, there may be times when you have to step up and be accountable even if the problems aren't of your own making. Most often this happens when taking over a system from someone else who didn't design, implement, or maintain it well. When this happens, there's a formula for establishing a timeframe to make things right before those hand-me-down problems start making you look bad too. The timeframe is usually about 50% *more* time than it took for the other guy to mishandle things in the first place.

Obviously, this formula isn't scientifically proven; it's just how things always seem to work out. If it was a bad job done by another tech over a two hour period, you have your own two hours plus one extra hour of leeway to make it right again before people start to see you as part of the problem, too. After that, anything that goes wrong is on you and not the other tech. Whining won't change this formula so don't bother trying. Just be accountable and work hard to make things right until the job is done. That's more than the other tech did before you took over so you'll still come out ahead even if you don't feel happy about it. This formula scales out to days, weeks, and even months for more complicated systems, but the math always remains the same.

Multiple system networks change this math somewhat by making it more calendar-based. There's no paradigm for measuring how long it takes to fix many broken systems at the same time. However, if you're taking over an entire network with big problems then the common rule of thumb is you have a total of nine months maximum to make everything right. Whether it's human nature responding to the seasons of the typical fiscal year or the time window for practicable performance, that amount of allowed time remains constraint in virtually every situation. After that, your coworkers will begin to suspect that the problems are coming from you and not your predecessor. His/her legacy will be your legacy at that point; therefore, if the problems are not entirely fixed with the accepted timeframe, you should at least have a clearly defined plan that demonstrates how and when things will be made right. You can use the plan to show progress to management and coworkers when they question how the repairs are going. You only have so much time to make things right, so if you're in a bad situation like this, don't waste time whining about it. Instead, it's best to just get to work and stay focused on the task until the appropriate results are finally achieved. Some ways to do this are:

- **Review and document everything in your own words** and don't rely on your predecessor's documentation. It was likely written using the same poor performance level that created the marginal system(s) in the first place. Redoing the old documentation not only helps to clarify how the system(s) should properly work, it also provides a better foundation to affect change. Doing this step creates a basis for further understanding of the company's overall operation. Digging in right away to work on understanding all aspects of the network from the start will allow you to modify items to your own liking much sooner than if you'd just sat and pouted.

- **Reverse engineer anything you don't understand**. Don't allow any left-over mysteries to linger around like ghosts in the machine. Just because you're able get something working the way the end-users are used to, doesn't mean you can fix it if it breaks again in the future. While reverse engineering can be tedious, especially when you're busy acclimating to your new job status, it's still better to do this as soon as possible. Otherwise, any changes you make to the system(s) could interact with your predecessor's legacy problems in ways you missed during the initial planning.

- **Prepare good arguments to give to management for funding any needed improvements**. When it comes to good arguments, the only one that matters is the one that involves saving money. If you can't cost justify the expense, then don't even

31

bother asking for it. The old adage, "if it ain't broke, don't fix it" will be applied every time by stingy management. If you know that the system is seriously dysfunctional and must be fixed immediately, still don't whine. In fact, don't even let a high note of frustration enter into your voice. Just go back to the drawing board and come up with a better...meaning more money oriented...argument to present to management. The importance of good communication skills will be mentioned frequently in this book, but suffice it to say, it's not your boss's fault if you weren't able to convince him the expense is worthwhile.

You must remember that a system's efficiency is one of the hardest things for companies to quantify. While they can easily add up the cost of every pen and box of staples in the office supply room, they cannot easily compute the financial cost of the difficulty involved with employees completing routine tasks with their computers. All those intangibles like morale, fatigue, and overall productivity must be clearly included in your argument when you're explaining the true value of your request to your boss. You also must be thoughtful with how you present the subject of spending money when making your case to management. You can't just toss in comments haphazardly when the boss isn't expecting them. Like it or not, there's a certain amount of salesmanship involved.

For instance, a good move when opening your argument is to use an old sales trick called Sign Posting. Salesmen have been using this trick for years. It's where they tell the buyer ahead of time how they're going to sell them something before they do it. It's meant to put the buyer at ease so they're less guarded when the salesman starts the pitch. For example a car dealer might say "I'm gonna sell you this car today and I'll tell you how I'm gonna do it: by showing you that we not only have the best cars here at So-and-So Motors, we have the best prices too." Of course you would use more appropriate wording for your work environment. The thing to remember is that if you're in the boss's office for a quick visit, passing them in the hallway, or even chatting in the cafeteria, you want to avoid ambushing them with a request out of the blue for something that'll cost money. Some planning ahead with sign-posting first will go a long way to not being rejected out of hand.

There are other approaches you can try too. Instead of using sign posting like a car salesman, you could also try something more subtle, perhaps some little hints you throw into conversations with management as an aside while you're discussing something completely different. Do it casually and smoothly enough every now and then and you may actually find them reminding you to pursue the upgrade as if it were their idea from the beginning. Regardless of which approach you use to sharpen your argument, be sure to always include the notion that the improvements being asked for will compensate for the costs.

At this point the whiner may complain, "Look, I'm a computer technician. Why should I have to learn about car salesmanship? It won't make me better at fixing computers." To which the reply is, "Get with the program." It's part of every IT professional's job, regardless of how tiresome it may be. In fact, in a world where every system in production today will be obsolete within the next few years, having the ability to sell upgrades to management is one of the most important skills you can have.

Whiners may also feel they aren't solely to blame for problems that arise. They may argue that it's the company's management that's causing all the issues. Well, even if management doesn't have a clue about what's going on or what a good solution should look like, it makes no difference. You're job as a professional is to get the systems you've been assigned to run at 100% operational efficiency by everyone's standards. If management isn't supportive enough in your efforts, then it's up to you to find a way to cut through the confusion, politics, or whatever, and get the job done anyway.

There's another approach that can also be used to get your way without whining. It can even work in companies that have unreasonable management. It's based on the fact that some managers respond to computer requests the same way they respond to new ideas...the only good ones are their own. It's important to develop the skills to deal with this type of manager because they are everywhere and won't go away. This's the type who always suggests the opposite of what you suggested. Perhaps they feel it makes them look knowledgeable when they dismiss the ideas of others out of hand. Nonetheless, this type of manager is easy to handle, but it takes some practice. The old trick here is to start off by suggesting the opposite of what you want and count on them to inevitably suggest the opposite of that. Sound crazy? Then you have no idea just how often this kind of thing actually helps techs get what they need to complete tasks they're handling. Spend enough time in an IT career and you're sure to experience this scenario at least once. Just be sure it doesn't backfire. They might accidentally agree with your first suggestion right off the bat and you end up with the opposite of what you wanted. Again, this takes some practice.

Another type of manager is one who is afraid your request will affect so much of the company, it could elevate you into a light that's brighter than their own. The trick here is to let them share the credit. Drop their name around as someone who came up with some of the original ideas. Sure it's not true, but you want those new systems right? So bite the bullet and get with the program. Chalk it up to the tough life of a computer tech…all pain and no glory. The bottom line here is that if you keep the systems in shape and up to date, your legacy will gradually come to the top anyway. The truth is, legacy is always better than moments in the spotlight. So don't let your pride get in the way of your good work. Rise up in your company through providing solid systems and not through craving attention. If giving up some credit in the short run means others will be commenting a year later (perhaps over the e-mail systems you built) on how good the computers seem to be running since you've be around, then you're on your way to achieving excellence in your field.

Yet another type of problematic manager is the "been there before" type. Whatever you suggest, they've seen it done differently at a company they used to work for. To them, that other way is better than what you suggest, regardless of how well designed your new system is. Now you're competing against something which is totally outside of your control, but this one is easy. Just remind them that their current work environment is different from their old company and that the solution you propose will work better and cost less in this situation than the one they remember will. On rare occasions they may actually consult a tech they know from their old company about your operations and how they can improve them for you. The problem here is that in the IT world no two systems are ever exactly the same and therefore second opinions are terrible ideas.

This's one of those few occasions where lowly techs get to say no to upper management. Unless the other tech is as up to speed on your systems as you are, then their ideas will probably do more harm than good.

Your systems are your domain and no outside intervention is allowed without your approval. If management lacks confidence in you, then improve both your knowledge and your communication skills. It's a different story if you requested the outside support on a professional level basis from a trusted vendor or through a manufacturer support agreement, but under no circumstances are you to turn over the keys to your systems to another tech who is behaving less than professionally. This's the one thing you should openly fight for with management. Just be sure to demonstrate good arguments without showing any frustration on your part. Once you finally get the improved solution in place and running well, it'll always strengthen your opinions in any future crisis. Just make sure those opinions never involve whining if you want them to really matter.

Too Busies

Even if your schedule really is overloaded, it's best to learn to stay ahead of the curve and keep the jobs you've been assigned on track. The most important tool for this is good time management. Few things are less forgivable in the fast paced 21st century than bad time management skills. In today's fast paced economy, time management is a critical part of nearly every job out there, IT or otherwise. Being able to manage your time is the same as being able to manage your usefulness. How much work you can properly do within a set timeframe is how your contributions to the company will be measured. In IT, this means managing multiple timelines while you move between the different jobs you're working on.

A major component of time management is the ability to prioritize each job as soon as it becomes known to you. Prioritizing means different things to different people. Some techs prioritize their tasks by focusing on end-user complaints. Others base priority on what they think will get the overall workload done the quickest. Still others prioritize based on where they'll be at any point in the day. Even local geography or a building's architecture can affect how you choose the order for getting things done.

Nevertheless, even if you're doing a great job with your own prioritizing, you can still be impeded by others who aren't as focused. However, to achieve excellence in IT you must always find a way to not let things slow you down. If the offending party is a boss who throws you a last second task, you still need to stay on track with all your original ongoing tasks. If he/she received a call from a customer that needs attention immediately, adjust your plans and add this job to the list and continue. Just stay flexible enough to be able to scale out your work load without stumbling to the point you end up getting nothing significant done that day, but don't go beyond your limits. The point when you become saturated by your workload and prioritization concerns should be well known and understood by you. If management doesn't know it also, then let them know. Your honestly and openness on an ongoing basis is a good thing and will make your boss's job easier too. After all, it's your boss who feels the most pressure from upper management and ultimately the end-users too. The more you can help them cope with the pressure by giving them honest feedback on the impact of the workloads you're being given, the better.

As far as end-users go, those affected by a computer problem usually feel their own needs are the most important. This's human nature, especially in pressure-filled work environments. It's just another variable for you to work through when prioritizing. When all the variables are working against you, don't panic or get frustrated. Just make good choices about how best to proceed and then stick to those choices.

There are a few things you can do to help manage time and make your work easier. They are:

- **Always respond quickly**

  The more you wait, the bigger the workload will be down the road. Today's jobs are likely to overlap with the ones you're sure to get tomorrow. Even if you're tired and worn out, quickly jumping on jobs as soon as they arise will serve you well in the long run. The more things you can cross off you list today, the better off you'll be tomorrow and in the future. Otherwise, you risk getting stuck in perpetual catch up mode. If several jobs come up at once, prioritize each appropriately but move on all of them as quickly as possible while remaining mistake free. Remember, habitually slow techs are easy to spot. They're the ones complaining about the work load even when they have less to do than other better producing techs around them.

- ## Don't call a job done too early

This's a recurring theme in this book. Calling a job done before it truly is done just adds to your workload down the road. Few things can hurt your morale more than having to stop in the middle of the day and redo a task you thought you had finished earlier. Priorities become confused and even the easy jobs feel a bit harder as you work to catch up. Doing a partial job only defers the core of the fix to some other time when your workload will be every bit as hectic. That kind of approach will always wear you out faster than doing a complete job does; especially after spending weeks, months, or even years perpetually catching up from behind.

Some techs try to make an art of balancing quick fixes like it's some kind of dance. In reality, they're more like the performer balancing a bunch of spinning plates. They move from job to job while hoping everything stays balanced and holds long enough for them to reach the endpoint. Unfortunately, this can only go on for so long before one plate eventually falls and causes the tech to loss focus; throwing the other plates out of balance, too. It's best to not get into this rush-job frame of mind in the first place. The tech shouldn't move on from any plate until each is confirmed to be stable and no longer at risk. Getting into the habit of sticking around long enough to honestly confirm a job is complete may seem a burden at first, but will definitely save you time over the course of your career.

- ## Handle time management now

Start planning for an easier future by including time management into the systems you're working on now. You do this is by performing the preemptive fix. If you're working on something and notice something else isn't quite right, do the second job too since you're already there. After all, that other problem isn't going to fix itself after you leave. If you ignore it in the hope you can sneak back later and handle it then, you'll only be wasting perfectly good time later on. Usually, with that approach the problem will just smolder until it eventually flares up before you return…and also when you least need it to. You'll allow the phenomenon called "putting out brush fires" to start taking over your work life. Leaving a tiny hot cinder of a problem unattended because it doesn't seem all that important at the moment will always cost you later on.

If you believe you'll get back to the problem at a more convenient date, you're likely fooling yourself. No one can predict the future, especially in the busy world of IT, so there's no way to really know exactly how much new work you'll be assigned tomorrow or the day after that. It's a gamble to think that a convenient chance to return to that proverbial "later" repair will even happen. In fact, things may be even worse later on if that shouldering cinder of a problem flares up when you're busy with another job that needs to be finished in a hurry, too. You'll always lose more time running back to put out brush fires than you would have spent had you corrected the problem in the first place. Best practice is to remember there's never a better time to make something right than the present.

Get used to inspecting the stuff you're working on and fixing all the issues you see even if they're not on the ticket. If it's a big problem that can't be fixed on the spot, you're at least ahead of the curve and can schedule a future repair in a reasonable non-brush fire fashion. The preemptive approach is an important and useful way to improve time management. Conscientiously working to stay ahead of the curve by nipping new problems in the bud makes the systems more reliable over time and your life a lot easier. Not only will you have less surprises, the systems themselves will be more bug free. Less bugs equals easier time management and happier end-users. It also helps you to establish a reputation for excellence in your field.

If you're running your own shop or are part of a development team, you have the power to be preemptive at the very beginning while in the design phase. Anything you can add in the beginning that will make solutions easier to fix later will make you a happier tech down the road. This isn't to say that things still can't get crazy with your work load, it just means the more you do up front in the design phase, the less you have to do later on. Every good design includes ease of follow-on maintenance.

No matter how busy you are, don't skimp on the front-end work for any system you're working on. Time management begins before you begin the work. Design your systems to be silent and trouble free. It pays dividends by giving you a system that's easier to maintain, causes less stress, and leaves you with more spare time on your hands. The old adage that you don't look busy because "you did it right in the first place" holds true here. Building things from the beginning with ease of maintenance in mind and then staying ahead of that maintenance makes task scheduling much easier over time.

It's a common misunderstanding by non-techies that the guys who are running around putting out the brush fires caused by poorly designed systems are hardworking heroes. Excellent techs know this is not true. Many brush fire guys are simply imitating a heroic effort while chasing after marginal systems that should have been designed better from the start.

- **Take brief notes outlining your work as you go**

  While this does mean doing paperwork, admittedly a painful subject to many techs, it rewards you with a good way to minimize your time on a job. This isn't about System Design Life Cycle, documented user requirements, IQ, OQ, PQ, or any other part of systems validation. This's about your own approach to keeping up with the things around you. It really can help to keeps notes as you work. Be sure to use plain words in simple sentences. Notes won't help much if they cause your eyes to glaze over when you try to read them back later. Techs do enough reading anyway. You should take just enough notes to help you from unnecessarily going over the same ground twice. Writing a few quick things down as you work means you won't have to reverse engineer what you just did when you go back to do it again at a later date. The notes are to help you keep track of what you're doing during a busy day.

This means labeling things too. Notes can only go so far when you're running around a busy job site. A good way to make things easier is to stick some simple but intuitive labels with useful info and tips on the items you're working on. This may seem like over-kill, but it's a good stress management strategy. The less you need to remember when you're rushing to fix something, the better your time will be spent. Especially have notes jotted down with quick tips for those systems you don't work on so often but usually need to be repaired in a hurry if they ever break.

Notes also help with new meta-directory items you add to network management directory structures. The same holds true for files and folders you create when working with file systems and storage. Always take a moment to add a helpful tip in the comment field or create a meaningful file/folder label you can understand at a glance. Your work will move faster in the future if you can see at a glance why a particular item was created. This helps your fellow techs too as they work on those same systems. The overall work atmosphere in your department will be benefit. After all, how your coworkers handle their time management affects your time management too, and vice versa, so the more everyone helps each other stay on track, the better for everyone as a whole.

Documenting is also a good way to be considerate to techs who follow behind you, too. When that inevitable time comes for you to leave your company for some other employment, the systems you leave behind will be easier for the new tech to take over. To any tech that doesn't care about how their old systems affect others, there's only one response: show some pride of ownership. Those systems reflect on you as an IT professional. If they fall apart as soon as you leave, it demonstrates your work only gave the appearance of quality and was never as good as it should have been in reality. Do everything in your career with a pride of ownership. This includes providing reasonable documentation for the next tech down the line to use. Take enough pride in your systems to make sure they keep running long after you've moved on. This not only helps you with job references in the future, it also helps you build a reputation over the length of your career as being a true craftsman.

Even if you don't plan to move on, good documentation is an important fallback if things ever go wrong. If you need to rebuild any part of a broken system, it would naturally be helpful to remember how it was originally put together. This is a tough lesson to learn. Trying to repair a system under pressure when you've forgotten how it was built in the first place is a bad thing for any professional to go through. You'll definitely lose some sleep over that one. Wise techs usually never let themselves do that a second time. If for no other reason than your sanity and self-preservation, have the main aspects of the systems you work on written down somewhere for future reference.

Even if you're working alone on your own network, keeping a knowledge base is always a good thing to do. This is especially true for any custom part of what might otherwise be a manufacturer default installation. Good documentation includes software keys, passwords, paths, licensing information, vendor lists, Service Level Agreement information, installation qualifications, and anything else needed to keep the system working well for a long time. Place this information in a location that's safe and accessible even in the event of a disaster.

If a solution was inherited from someone else who didn't do the best job writing things down, then do the math to determine how much time you need to invest to make things right. A good rule of thumb is that reverse engineering a well-built system to make it better understood, more easily maintained, and above all, more stress free, takes about the same amount effort as it took to design the system from scratch. Create the documentation just as with everything else you designed and work on. Don't let some other tech's lack of professionalism affect your own. Even if it wasn't your solution, you should still put your pride of ownership into every aspect of it. Every excellent tech always does.

Experimenters

Rule number one for anyone experimenting on production systems is to stop immediately and don't try that again. Especially if you don't fully understand all the potential interoperability impacts within your company's operations. If you don't know the difference between a safe place to experiment versus the systems end-users depend on, then you're a liability waiting to happen. Whether you really need to learn how something works or are just the curious type, slow down and have the patience to build a learning environment that's not attached to anything important. Something like this may already exist if your boss understands that techs never stop learning. If not, you or your boss should either find a way to keep old equipment around or buy appropriately scaled equipment for techs to use when playing around with a solution they're still learning. Using some spare space to build a reasonable computer lab where techs can train in the workplace pays big dividends. This isn't to be confused with a pilot environment, which is far more structured and monitored, but instead is a room where things can be built, torn down, and built again as part of an IT professional's learning experience.

If a learning system can't be built without extra licensing or proprietary hardware, then make this part of the vendor selection to begin with. If a vendor wants to ensure that their systems look good in production, then they should encourage professionals to practice on their systems continually. This means providing as much hands-on access to their system as possible without giving it away for free. The vendor may offer first-rate tech support, but that's usually only after something goes wrong. At that point, the system they sold you may already look like a bad investment no matter how well designed it was. The same is true for installation support that doesn't allow for a productive pilot to be completed first. After all, your computer environment is not, no matter how default it may seem, identical to any other computer environment in the world.

A vendor looking bad means their product gives the impression it needs a measure of luck to remain at least 99.9% operational. They aren't allowing for the reality that luck can never be a factor in uptime. As long as the vendor makes it difficult for you to continually train on their system, there'll always be at least a tiny bit of luck involved when adding something new to their solution, no matter how careful you are. It may be better in the long run not to purchase systems from vendors that fit this mold. The ability to test and continually learn on new products, add-ons, or systems should be a routine requirement in all your vendor procurement cycles. Ongoing learning ensures the system will consistently work well in any situation. If a vendor's product offers consistent performance, their brand will speak for itself and carry more weight in the industry.

Regardless of whether or not you have a learning lab in your workplace, always learn in an environment that doesn't touch the production systems. However, this environment must resemble the production environment closely enough to produce useable results. The idea is to design the lab so that you can create as little stress as possible when you move your newfound knowledge into production. The more you know ahead of time, the more at ease you'll be when the work finally begins. You'll also look better organized since you're cool-headed and ready to go when implementation begins.

Of course, no matter how much understanding you have at the beginning, things can still occasionally go wrong in even the best environments. Setbacks can be caused by any number of factors outside your control; two examples are a regional power outage or a bit of undetected malware in a connected system. The important thing to remember is that fixing something and learning how it works should never go hand-in-hand. Don't allow bugs in the system to be your tutor. Knowledge isn't something best gained during a disaster. It must always be there a head of time. It's never enough to just install a solution and get it running well; you must also think ahead and understand all the ways it can break down the road. Try to test as many of the problems that can defeat your solution as possible during your training. When it's completely broken, get it running again. The more you do at this stage, the more prepared you'll be for whatever fate throws at you later on.

If something ever does go wrong in a way you hadn't thought of earlier, be patient while you work to get it repaired. Use your communication skills to keep management from pushing you to move too quickly before learning what the best fix is. If this means a quick trip back to the lab, so be it. It's better to suffer through clear and measureable downtime than to quickly have the company running on a marginally restored system which includes hidden costs that are not easily measured. Your credibility will suffer more in the long run from building solutions that constantly need small repairs than it will from overcoming one big outage successfully, even if the second option took longer to get done. Credibility is a day-to-day thing. It's easier overcome a delay that resulted in a successful repair than it is to overcome persistent small problems that serve to remind everyone your work may not be the best.

Constant Rookies

An all too common way to avoid stressful situations is to avoid taking on the tough jobs in the first place. In most walks of life, there's no easier way to avoid hard work than playing dumb. Not only can playing dumb get you out of doing the occasional tough job, it might even make management think you're a reasonable employee for being so "honest" about your lack of abilities. After all, letting them know ahead of time you don't feel you have enough knowledge to do a proper job may make you look downright conscientious, never mind if your grumbling coworkers know better.

The problem with this style of career management is that it only goes so far before it backfires. The slow pace you've set for yourself by backing away from tough jobs can lead you into falling behind on the critical task of learning new things. If you get too used to hiding behind your coworkers as a form of stress relief, the day may come when you find yourself so far behind you're no longer employable. It's better to accept that all rookies must one day accept the mantel of the experienced professional. When that day comes for you, step forward no matter how tough the job seems to be.

The problem here is that if you get too good at hanging back, you might find out you've forgotten how to step up when you finally decide to. What's more, your boss may just quit calling on you out of habit whenever a new job appears. After all, you've always backed out before. When one day you finally decide you're ready to get into the game and put forth a real effort, you may find yourself left too far out in the cold to participate any more. That's not fatal to your career; just convince your boss you're ready for more tasks. However, it's an embarrassing point in your career that you may not live down so quickly.

The worst case scenario is to find yourself playing catch-up when your job is on the line. One of the first requirements of a company struggling to stay afloat in a competitive industry is to have aggressive and capable employees on staff; the more self-starters on the payroll the better. If you find that your employer is beginning to pursue employee cutbacks for whatever reason, you're more likely to be on the cutback list if you're a persistent underachiever. It's a fast fact and can't be undone by last minute appeals or a sudden surge in productivity.

Your long term track record says more about you than anything else. Nothing you do in the short term can change it overnight. It's best to develop a good work ethic to the point where it becomes second nature. If you build the reputation of doing jobs well and on time, then the rest of your career will take care of itself regardless of any company's particular economic issues. Even if you do get laid off, you'll likely have good references to take with you on your search for that next IT position.

Divers

Divers are a lot like Experimenters, except they're more about cutting corners than learning as they go. Divers are smart enough to know better than to dive too quickly into a job without completely understanding it, but they dive in anyway. The thing about divers is they were once good techs who did good work, but now are getting sloppy at it. They should know better than to dive into anything without knowing what's beneath the surface, but they've done so much good work in the past they've convinced themselves there are no surprises anymore. They think that since they've made it through this type of thing before, they'll make it through again.

If you think success can be measured merely by counting your past laurels then you're setting yourself up for failure; if not this time, then the next time or the time after that. Make no mistake, if you're too good to bother trying anymore, then failure is waiting for you around the next corner. It's a strange truth that few things can cripple an accomplished person more than continued success. The Japanese used to call it "Victory Disease." It's overconfidence to a fault. Just because many battles have been won in the past doesn't mean more will be won in the future without trying.

The first step toward curing yourself of the Diver affliction is to make sure you always understand that as a professional, you should never let fate to teach you life lessons. That's best left to beginners just starting out or to professionals who have lost interest in what they do for a living. To control your own destiny, you must never allow yourself to believe the road you travel will get smoother as time passes by. Computers just aren't that accommodating. They'll always be ready to trip you up you no matter how good you are.

Excellent IT professionals know there'll always be more bumps ahead; after all, bumps are a part of life. More importantly, an excellent tech never forgets that no amount of prestige and experience can make egg on their face look good. Before you continue down that road, learn to stop and think carefully about any job you're about to undertake. Make sure you take the time to understand all aspects. Do this the same as if you're a beginner. Don't ever become convinced of your invulnerability because in reality, it will only make you vulnerable.

Besides, how much fun is living life without challenges...really? If you think you can skimp on the details because they'll be pretty much the same as before, what you're really skimping on is the challenge. This isn't a small thing. Being able to step up to challenges is the most important part of your career and the biggest challenge will always be found in a job's smallest details. Never ignore details. It can be said this way; small details are the brushes that you as an IT professional paint your craft with. There's nothing prettier than a well thought-out project that brings some important system into production completely trouble free. Even a thoroughly complex system with all its moving parts and complicated add-ons is a thing of eloquent beauty when all its small details fall cleanly into one piece without a hitch.

Divers however risk losing the ability to continue doing that level of work. Their growing habit of cutting corners will soon leave too many details ignored for them to remain effective in their craft anymore. Even if they're convinced they've seen it all before, the truth is that knowledge and ability don't work that way. Divers need to still care enough about what they do for a living to always find the time to slow down or wait until they're truly ready to start a new project.

Know-it-alls

Know-it-alls aren't usually bad, it's only when they start to replace knowing with pretending to know. The truth is, expertise holds little value in a world where skill-sets turn over every three years or so. In many cases, by the time you've become an expert on anything in IT, it's already starting to become obsolete. To stay successful as an IT professional, it's not how much you know but how fast you can keep up with learning new things that matters the most. This means that while cracking those proverbial books will always help you on the job, what you learn from those books will only take you so far before it loses its value.

The problem is that some techs hold on to old knowledge too long after its usefulness has faded away. Since they've already experienced the power and respect that expertise brings, they can't bear the thought of going back to the beginning and starting all over again on something new. It's much more reassuring to rest on their laurels. However, they forget the hard truth that underlies all careers in information technology: when your particular bit of expertise becomes obsolete, it's back to square one with all the other rookies. There's no way around it. A know-it-all's unwillingness to acquiesce to this truth can lead to major issues for their company.

Few problems can be bigger than a know-it-all who, after obtaining great knowledge on one system, leverages their well-earned respect to give bad advice on another system they don't understand as well. That kind of thing can be a back breaker for everybody involved: the team members who have to go back to the drawing board, the manager who is now way behind schedule, and especially the know-it-all who just proved to everybody that they really don't. To leverage out of date respect into leading others to trouble is an unforgivable offense in IT. It amounts to an act of sabotage, no matter how well intended the effort was. Careers can take a long time to recover from that kind of faux pas. It's best to resign yourself to the idea that as a good tech, you'll inevitably spend your whole career climbing to the top of Mount Olympus only to roll back down to the bottom again every time the industry heads in a new direction.

The most successful approach to this never ending struggle is a sustainable one. More than anything else, the most important tools of your craft are the learning methods that work the best for you. Anyone can be a quick learner once they find the formula that works best for them. Whether structured schooling or self-study based, finding the method which helps you learn the easiest is the biggest step in your career. Until you take that step, you'll always be behind the rest. If you miss being the most knowledgeable tech around each time something new hits the market, just remember how you became the most knowledgeable tech around the last time something new hit the market: by learning more about it than everyone else did. This means finding the easiest, most stress free method of learning that works for you.

Some techs like attending learning boot-camps, others go back to the more traditional classroom, and still others practice the discipline of independent study because it fits both their budget and their schedule. The thing to keep in mind is that if you start to tire of chasing new knowledge, don't try to replace it with false knowledge that only carries weight because of the name you made for yourself in the past. Other techs might still be listening to you long after your knowledge has lost its accuracy and efficacy and, as a result, will suffer from system issues caused by your misguidance. If you're truly a professional, then take the time to learn what you need to know before sharing any knowledge with others. Excellent techs don't allow themselves to pretend to be knowledgeable in a world where knowledge is fleeting; they make sure they're knowledgeable every day they go to work.

You must resolve yourself to the cold hard fact that learning and obsolescence will continue to haunt you throughout your career. It causes more career burn-out than anything else. Learning brand new things every few years may seem like an adventurous challenge to someone starting out in IT, but even the most hardened professional can begin to dread the effort it takes to start the climb back up the knowledge mountain all over again with every new release.

This's the life of an IT tech. Sure, some of the core fundamentals like network protocols and programming language have been around for a long time, but it's what's happening around them that you need to especially keep track of. That's where you constantly need to develop new expertise. What's more, even those core fundamentals will change eventually. The sad truth for know-it-alls is that no one can know it all forever. Never pretend to continue to have expertise after you've burned out and given up the chase. If you're ever afraid to take on a new IT project because you don't want others to see how far behind you've fallen in your skills, then it's time to consider moving on to different job challenges and perhaps even a career in another field.

Fraidy-cats

Fraidy-cats are rookie techs that want to work, but are bit too insecure to take on big jobs when the time arrives. This's especially true if they feel other techs around them know more than they do about the same system. If this describes you, then you must keep in mind you're part of a team, and as such, you have a duty to carry your fair share of the workload. This remains the case even when you're not sure how well you can handle the load.

Same as with any job that requires you carry big loads, you can benefit greatly from exercise. This means that anything you can do to build up your load bearing capacity will help your skills and your abilities grow. If you've ever worked in a warehouse or on a construction site during your life travels, then you already know that the more strength and stamina people have, the easier it is for them to move things around. Lifting weights after-hours to build even more strength is a great way to improve personal performance in those particular career fields.

The same goes when working with computers. Even though the effort is more mental than physical, exercise is still part of the game. The stronger you become with computers, the more your confidence will grow. To begin with, don't wait until you're assigned to work on a particular system before you start learning about it. As soon as you hear the system is under review for possible implementation into production, you should start reading up. Better still, volunteer to be part of the product evaluation process. That way you'll definitely know more about it than most others around you when it's finally deployed to the end-users.

The biggest cause of Fraidy-cat syndrome is waiting to learn about something new until the knowledge is too late to do you any good. You'll be left feeling behind the curve and not quite ready as the move to production phase approaches. In all fairness, you should feel this way. You've under-prepared in a rapidly changing industry by being less motivated in your product knowledge and review than the other techs around you. Remember that except for the most basic networking core protocols and practices, almost everything you learned in computer school will be long obsolete before you gain even mid-level seniority at your very first job. When the day finally arrives when the boss decides to give you your first IT project to run, you'd better not plan on falling back on the book learning you received back in school because virtually all of it'll be out of date. You'll need to have a whole new set of personally motivated book learning initiatives to fall back on. Whether you obtain the new learning from more college courses, expensive boot-camps, or, most commonly, self-study, it's what's going to help you to get the project done right. Bottom line: the more you know, the more courage you'll have at the start and the more confident you'll be along the way.

Another thing to keep in mind: don't let a snooty co-worker or office politics eat away at your ability to believe you're the best at what you do. No matter how bad it may seem at times, the truth is the only real judge of how good you are with computers are the computers themselves. Keep them running, upgraded, and migrated on time and your confidence will grow. Uptime is the one thing that no one can take away from you. Uptime is the ultimate compliment and the best reference you can ever have in the field of information technology. There's no better IT professional than the one who keeps everything they're assigned to work on well maintained and running smoothly. In fact, that's pretty much the only kind of tech your boss ever needs. Get on top of the systems you're working on and any Fraidy-cat feelings you have will subside in no time.

Nerds

Nerds tend to neglect the communication skills necessary to gain the status of IT excellence. This's caused by their lack of empathy with the end-users out there on the front end of the systems the nerd has been assigned to work on. Depending on how cynical you are, you can view a career in computers as either an adventure in sociology or an adventure in babysitting. You're like a high priest of mystery and magic: you carry deep knowledge of an invisible place where everyone around you goes to do their work, but only you can navigate. This causes people to view you with both fear and envy, as well as both respect and dismissive contempt. How you handle your role as both the creator and guide of this mysterious place called computers is where those communication skills come in.

While you don't need to speak out loud to computers to make them work for you (at least not in this human generation), they still hear you talking anyway. If fact, computer systems hang on your every input like it's manna from heaven. Keyboard strokes and mouse clicks are their voice. Nevertheless, you must remember that it's not just a matter of how gifted you are at speaking with your systems. You must master the art of talking with your coworkers and business partners, too.

Remember that even if you're the only one in your department, you're never alone. If you run the network entirely by yourself, you still need to interact with day-to-day vendors, service level vendors, support techs of all shapes and sizes, end-users, and the ever present upper management types looking for ROI (which means either

Return on Investment or "Reason of Insanity" depending on the boss's techie rating). If you work in IT, your communication skills are a critical part of your career. Be sure to keep an eye on them. This applies equally from the "Smiling Jacks" who can talk themselves out of any disaster to the reclusive geeks who are all business because "humans are too much trouble anyway." Communications skills must be developed and maintained throughout every IT professional's career. Without these skills, both you and your computers will suffer.

This's because you're the translator, the interpreter that stands between your computers and the world of the living. As a tech, you have the ability to speak with something that only exists in a virtual haze and allow it to properly interact with the flesh and bone people in your company. That means being able to understand the requirements the end-users have and then working with the computers to bring those requirements into production. For this relationship to function successfully, you need to be able to communicate the needs of computer users to the systems you're working on. Without good communication skills, maintaining a high level of functionality with the systems you design and/or maintain will always be a tough task.

So what are good communication skills? Well, to begin with, less is more. Remember that great truth that there's no better measure of how well somebody understands something complicated than how simply they can explain it to the average layman. If you can use analogies, mental pictures, or any other holistic approaches, then do it. For example, explaining to a non-techie that an SMTP transmission is like a frog leaping across a pond from lily pad to lily pad is better than confusing them with specs on store and forward retention times. Use your best judgment in each case. Even if staff is legitimately concerned with something like e-mail privacy, responding with too much detail could be as ineffective as not responding at all. Confusing them with specs will not alleviate their concerns. A more balanced explanation that perhaps mentions frogs and lily pads may be a better choice in the long run.

Sure, this may seem a bit demeaning, especially to those in your network who are confident they know as much about computers as you do. It makes no difference. Never forget that unless people build and maintain computer systems for living, they don't know anywhere near as much about computers as you do. This's critical because their lack of true knowledge combined with their overconfidence can result in them making assumptions about what you're explaining to them that are totally inaccurate. They may hear your words but not in the context you intended.

It's never good when, after a new system goes online, someone complains that they thought it would have some new-fangled feature they read about in a blog somewhere. The truth is you promised no such thing, but you allowed this misconception to happen. You gave the benefit of the doubt to how well that person understood your original explanation when in fact they were drawing on their own limited understanding while working through your overly complicated details.

If you get in trouble for this kind of misunderstanding, it's always your fault and not the end-user's. This's as it should be. Use the experience as a reminder that you need to work on your communication skills. You should always keep it simple, even for the power users you think might understand what you're saying. Never feel you're talking down too much to anyone's intelligence in IT. Everyone appreciates an accurate, clear and simple explanation of how something complicated works. It allows them to get back to all that complicated knowledge their own busy careers require. Above all, it ensures the ROI exactly matches what they thought it would from the start.

Another thing to remember about communication is to always stay involved when you are on a team. If you don't like a project, or can't commit your full effort to it for any reason, tell the team leader at the start. An absentee team member is a burden to everyone else. Absentee in this instance means anyone on the team who isn't participating fully. Actually, even if you attend every meeting and perform most of your assigned jobs on time, if you aren't communicating your input well to team members, then everyone else will have trouble staying on track. Good communications are especially critical if you made a mistake somewhere. Admitting to it quickly is a good measure of how committed an IT professional you truly are. Every professional makes the occasional mistake. Leaving your mistakes for others to discover makes you the silent partner nobody needs. In other words, you're nothing more than an absentee team member if your communications are not complete, current, and clear. If you are involved on the job, then part of that involvement is to communicate well and fully with everyone.

Breathless Wonders

It's not merely a coincidence that bad time management skills and panic go together, but sometimes panic isn't about a lack of time at all. Constantly demonstrating a panicked state because of an "incredible" work load may simply be a way to look more important. Appearing constantly overwhelmed by your work load is definitely not a good way to establish credibility, although some techs try to do it anyway. It may buy you some immediate sympathy if you're feeling a bit ignored, but it can also become tiresome for those you work with. In fact, the timing of your next panic attack might even be an office bet in more cynical IT departments. This's not the way to build a reputation for excellence.

Worse still, your panic attacks could become contagious. If another like-minded tech becomes a bit envious of the sympathy your panic attacks get you, they might start touting their overwhelming work load too. Pretty soon you have an office burdened by multiple techs who complain every time a big job comes up. What's more, this can badly affect moral. Breathless types have a way of demeaning others around them by inferring that everyone else's workload is less than their own. No matter how busy everyone may be, they'll always find a way to *sound* busier, true or not. This can cause other IT professionals around them to feel unappreciated despite the good work they're doing.

The real risk though is that the tech who acts overwhelmed all the time might eventually be listened to by management. That means less work coming their way over time. This may seem like a good thing at first, but before long it'll be clear to everyone in the department that the breathless tech no longer has a workload worth complaining about. That puts the Breathless Wonder in a negative situation because their principal means of getting attention will disappear. When that eventual point finally arrives, they'll just be another underachiever taking up space.

Don't ever let yourself get into this situation. If you aspire to have a legacy of excellence in your career, going the Breathless route isn't part of the game. It's far better to bring attention to yourself by doing good work with every task you've been assigned. The best part of doing good work is that if you ever really are overwhelmed at some point, others will take you seriously and provide assistance when you actually need it. There's no shame in this. As long as the computers around you are running smoothly all the time, there's nothing at all wrong with getting a little help every once in a while to keep the uptime going. Just don't sound rushed and panicky when you ask for help. That's a road best not gone down…especially if establishing solid credibility is your long term goal.

Finally…

If you find yourself falling into any of the aforementioned tech types, or a blend of those types, you should begin the process of evaluating how good you are at your job and find ways to do it better. The path to excellence is the path to working happier and longer in your career Even if you settled on a career in IT rather than planned for it from the beginning, you'll be amazed at how engaging and fulfilling it can be if you use the right approach. Having a job that starts everyday afresh with new sets of challenges is both exciting and rewarding if you handle it well.

CHAPTER THREE

Stress Management

The first thing to plan for with every new job is how you're going to handle the stress it brings. This isn't meant to be cute, it's really the truth. One of the truest measures of how well a job was done is how little sleep you lost along the way. The math is simple: no lost sleep equals a great job done. Lots of lost sleep means it's time to get better at what you do for a living. It really is worth the trouble. Nobody should lose sleep over their work, regardless of what they do for a living.

No matter how resilient you may think you are, stress is sure to wear you down to nothing over time. Before you know it, even the most routine tasks you've enjoyed doing in the past start to become exhausting. Things which were once easy to do will begin to feel beyond your reach. The effort needed for even simple jobs seems to keep increasing as you get older too, but don't fool yourself into believing it's simply the result of aging. In fact, it's wrong to even consider age as part of the problem. Modern science shows that if you treat it right, the human brain never really gets old. Nonetheless, the brain can definitely be stressed into inactivity and even chronic incompetence if enough stress is applied. If you want a long, happy career in a pressure filled world working on complicated and important things, it's best to figure out early on how to keep stress out of the picture.

Stress in not synonymous with words like complicated, difficult, and pressure. To believe that is to be less than motivated than you should be about the impact stress can have on your career. Excuses abound in the attempt to make stress look like a normal part of the job. For example: the work order came at the last second, the user requirements kept changing, the functionality some in the company are dreaming about is nowhere near to what the company can afford, or, of course, the work load is too great. Excuses are common everywhere you go and you'll hear them all the time, but in the end none really matter. The truth is that if you ever lost sleep over a job, it's only because you failed in your primary understanding of how to keep stress out of the picture. It's a harsh truth that your computer systems will never let you forget.

Stress management is one of the most practical skills you can carry with you. It affects how well you perform with everything in your career; from learning new material to troubleshooting tough problems. Even your ability to make a good call about the direction your career is heading can be impacted by how well you're managing stress at any given time. Stress will follow you when you head for home at the end of the work day. It will eat at you in traffic, at the dinner table, and eventually, when you're trying to sleep at night. The sooner you understand how stress works, the sooner you'll be able to mitigate the harm it's doing to your well-being.

Start by remembering that the amount of stress you experience when doing anything is directly proportional to the lack of experience or understanding you had at the beginning. Stress is the result of not having a clear perception of the future when you're first starting out on something. This applies to all things in the world. Whether it's an important job, a tough exam, a brave repair attempt, an act of gracious volunteerism, or even a well-deserved vacation, if the beginning expectations were unrealistic from the start, then things won't turn out as planned. Anytime something doesn't work out as planned, even if it turned out better than originally hoped, a measure of stress occurred somewhere along the way.

This means that the best time to practice stress aversion is before you start off on a new path. Best practice is to form a plan on how to do something in your most optimistic frame of mind while taking into account the expectations of everyone around you, then destroy that something in every way you can possibly imagine...no matter how improbable those ways may be. Sound cynical? Call it hard learned experience. If your approach to a job is only formed by your desire to see smiling faces and sighs of thankful reassurance when you're done, then you're setting yourself up for stress. Always take full stock of a situation before promising anything and continue to remember that no truth is better than hard truth, because no other kind of truth gets the job done right.

There are two things that you need to do well as a tech and both take place when you're starting a new project. One is the ability to correctly estimate the costs needed to complete the project and the other is to correctly estimate the amount of time required to do it. If you start off by low-balling the cost or time schedule to make others happy, you risk them being frustrated when things don't turn out as planned. Their frustration can be prime source of stress for you. On the other hand, if you don't overstate the initial cost and time required and then finish the task ahead of time and under budget, you'll look like an IT professional who gets things done right.

You'll learn over time what costs estimates work the best for nearly every situation by paying attention to how the procurement of materials needed for the job are applied to the things you do. This comes from just remembering how much stuff generally costs. However, learning to estimate the time required to complete a job, especially a large job, is a bit more difficult. Every activity in IT is under some time constraint as a matter of default, so how well you manage your time and effort will ultimately be the final measure of how good you are at your trade. The work needs to be done and done well within the anticipated timeframe. Things will go badly if the original plan didn't take into account enough time allowance to provide for things that can interrupt the work flow. Always be brutally realistic in your time estimates too; not just for your own sake, but for the sake of those impatient management types as well. They have their own agendas that need attending to same as yours. Everyone should be on the same page about the time needed *before* you begin a job or stress is sure to follow.

It helps to give due consideration to what the requester wants in terms of a schedule, but then only agree to the schedule after adding in a reasonable fudge factor. The following formula isn't meant to be funny; it actually works in the real world. Give it a try and you'll see.

When it comes to big jobs, the fudge factor is the anticipated length of the project plus an additional: 15% to account for the waffling of outside parties, an additional 10% to get everybody REALLY up to speed, another 10% to work around the fact that some team members will never be entirely on board, an extra 10% to explain to the boss why you're still doing things exactly the way he originally wanted when there's a "better way" he discovered after you began, plus 10% to tweak things to convince the boss the best ideas were his even if he had nothing to do with them, and finally, 20% more time to cover any remaining wrinkles when moving to production phase…i.e. problematic pilot groups, etc. In fact, forget the math and just double the requester's original time estimate, add a little extra for good measure, and you'll have a workable job schedule. Get good at this kind of calculation and your life will be easier and more stress free. This math even holds true for emergency repairs because emergency response always benefits from the same level of thoughtful planning as dedicated projects do. No matter how critical the task may be, keeping expectations for its completion within a workable timeframe keeps everyone happier than simply planning to get things done "ASAP".

Keep in mind as you go through your career there are other things besides cost and time estimates that will bring stress to your job on a daily basis. Being prepared for these things and finding ways to avoid them will give you much satisfaction with the work you do. Everything is be easier and more accessible if you can keep stress out of your daily routine. In particular, be mindful of seven important causes of work related stress you're sure to experience at some point in your career…everybody does. You should manage these stress builders every day to ensure whatever career path you follow in IT will leave you feeling fulfilled and satisfied.

Seven Causes of Stress

There are seven causes of stress that can affect you on a daily basis when working in a computer career. They are:

- Using the wrong approach
- End-users
- Overreaching your limits
- Not delegating properly
- Delegating without checking first
- Forgetting to delegate in the first place
- Lack of motivation

Using the Wrong Approach

Whether it's implementing a new solution, upgrading an existing solution, or repairing something that's broken, there are two wrong approaches which should never be used. They are working too quickly and working too slowly. Both will cause you to lose sleep at night. We'll cover working too quickly first since it's the lesser of the two evils.

Working too quickly and losing track of details is probably the most common mistake techs make on an everyday basis. They forget that fixing something just good enough to get by is really the same as not fixing it at all. Both will leave the complete fix unresolved...yes, unresolved to different degrees, but unresolved to some measure nonetheless. This kind of job related behavior is common, too common in fact, for an understandable reason.

Doing a quick fix and sneaking away is easier than spending precious minutes making sure everything is working perfectly before going on to the next thing on the list. It's safe to say that every IT professional has done a quick fix at least once in their career. Typically, it happens when working on a system for the first time and feeling overwhelmed by new technology that the tech didn't take enough time to learn about. A solution is quickly patched together that's good enough to get the system working for the time being and the they quietly head out the door hoping for the best. The focus is limited to short term results while the long term implications are ignored for convenience sake. This remains true even if the "quickly patched together" act takes several weeks of effort to accomplish.

If you ever find yourself guilty of performing a quick fix, you need to stop and ask yourself why. The quick fix isn't just a bad habit, it's an addictive habit. Take a close look at your situation and try to remember how and when you got into the quick fix frame of mind in the first place. Be truthful in your self-assessment here. Don't just blame the other guy, especially if the other guy is a boss who's giving you too much work to do in a reasonable amount of time. You might be surprised at how easy it is to be your own worst enemy when it comes to the being thorough in the work you do.

IT departments face the same economic pressures as any other department in the company. Response to that pressure is often downsized payrolls, hopefully consisting of employees who can do their work effectively. In poorly managed companies, this can lead to overwork for remaining staff, which ultimately hurts overall efficiency. Overwork then leads to more inefficiency so eventually there's a demand to hire more people to pick up the slack. It's a tricky cycle to navigate, trying to balance departmental budgets with departmental output. As a result, upper management has to find a balance that allows for profitable output from the fewest individuals while still providing a work environment that can remain productive over a long period of time.

There are ways you can help with this balance that both lowers your stress level and the stress level of upper management too. First, take a good look at yourself. Can you work more efficiently? Are there skills you can hone that'll help you design or diagnose a solution more quickly? A little knowledge can go a long way toward removing stress from your life. The more you know about a system you're working on, the more stress free you'll be. This means you can work faster while still avoiding the quick fix. Working while only guessing how things should proceed is a major stress maker. Guessing leads to doubt and nervousness which are both at the very heart of stress. If you're working as efficiently as you can, you should be moving through each day stress free no matter how busy you are. Efficiency means understanding your limits. If you're suffering under a work load you volunteered for, then you need to tone down your ambition a bit or you'll burn yourself out.

However, if the oversized workload was handed to you by a poorly managed department, then take a solid assessment of the situation. Making decisions which have the potential to affect your long term career isn't something you should do when too harried to think clearly, so find a calm space to mull things over. Be sure to look after the most important asset you can bring to any job: your professionalism. Overwork can lead to mistakes which could damage your long term credibility.

Go back to your communications skills and use them to convey a clear message to management that an untenable working environment is an uneconomical one too. Overworked employees turn into unproductive employees and an unproductive workforce is worse than no work force at all. Employers don't have to pay wages if they have no workforce. However, they must continue to pay unproductive employees even if the wages don't bring in the returns they hoped for.

Providing this feedback to management helps them to find a better balance between wages and output and may bring you the relief you need. Handling your workload well regardless of the situation will define your path to excellence. If you maintain good skill-sets and manage your time well enough that you're not leaving a trail of quick fixes, then you might be surprised at just how much work you can get done. You'll find yourself working very quickly, but not too quickly. What's more, every job you take on will be 100% complete when it's done. When you're handed what appears to be an overwhelming work load, take a closer look to confirm that it really is overwhelming. You may find that good work habits such as regular learning and consistent documentation can help you shrink that workload down to a reasonable size.

The other side of the problematic work flow equation is working too slowly. Working too slowly may not seem bad at all since stress levels will be comfortably low because of the nicely relaxed pace. After all, if someone wants to avoid stress, working as slowly as possible is a good way to do it. The problem is that as deadlines start getting close, working too slowly can quickly turn into a big stress generator. Stress can come from frustrated clients, angry bosses, or even other team members whose jobs you might be impeding. Working slowly may feel easier at first, but what you're really doing is deferring the work until a later time which is certain to hectic when it finally arrives. The worst case scenario is you'll lose credibility because teammates have to step in and help you catch up with your workload.

The only cure here is to plan well and move into each job in an assertive and confident manner. Don't be risky, but don't get stalled by contemplation either. That just tells others you may not be as confident as you should be. If you're worried about being over your head on a particular job, then deal with the worry head on. Bolster your confidence with study and practice until you're comfortable with the task at hand. Rehearsal can range from trying things out on test systems to reviewing repair steps in your head while traveling to a job site. While it's good if fear holds you back from biting off more than you can chew when volunteering for jobs you're not quite qualified for, don't let it stop you from moving forward in reasonable steps. Just be careful about how you proceed and you'll be able to balance ambitious risk-taking with thoughtful stress management every time. The thing to remember is that slowness leads to stress and finding ways to work more efficiently and in a timely manner is a positive step toward reducing stress in the long run.

Working too slowly is also the result of allowing someone less knowledgeable than you to cause you to suffer indecision. It's part of your job to communicate well enough to instill confidence in those around you. Once you decide on a path to follow for a job, you must stick with it. Even if a meddling boss comes up with another approach he just heard about, find a way to convince him that your first plan of attack is still the best one. Never second guess your first notions when you starting off on a new task. Allowing others to over think your solution is the quickest way to bog it down. While the oft-cited acronym KISSASS, or, "Keep It Simple Stupid, it's Always Something Simple" may not apply so well when working with all the wonderful details that make up a beautifully engineered system, it does help to remind you that intricate systems are complicated enough without you adding to that complexity by thinking in circles.

The same goes when getting input from others. Don't allow others to over think your project either. Once a path is set, stick to it and don't deviate until the job is done. Do this and you'll sleep better at night, not to mention work faster too. If you did your planning correctly at that beginning and started the job fully prepared, you'll be right every time anyway.

There a few other things you should focus on when dealing with the bad habit of working too slowly. Following is a list of eight points to keep an eye on:

- **Poor organizational skills**

  The good news is that having poor organizational skills is and easy problem to fix and once fixed produces immediate benefits. Good organizational skills make everything go smoother. You'll enjoy a blissful combination of less stress and more productivity. What's more, your boss and your teammates will notice how everything seems to go better with you around.

  Organization should be the first thing you do at the start of every workday. Write down your chores for the day and update an overall list of your big tasks that you should also be keeping. Lists are a great way to put your work into perspective. Even when planned jobs are interrupted by a rush assignments out of the blue, you can use the list to get yourself back on track later on. Making a list may seem tedious at first, but only takes a few seconds of scratching some quick notes onto a notepad each morning. Think of it as a shopping list of tasks you want to accomplish for the day. For the month list, use whatever text editor you have handy and create a simple list with expected completion dates, then update the list on a regular basis. You may even set it to launch to your computer's desktop with your boot up every morning. Keep in mind no one else needs to see it. It's only for your own purposes. Lists help you move faster throughout the day by removing the need to constantly stop and take stock of what you're doing at each step. All you need to do is look at the list and you're on the path to the next task.

A daily task list also goes a long way toward minimizing down time when moving from one job to the next and therefore helps you maintain a better level of overall performance. Trying to decide what to do next is a real time waster in the multi-tasking IT world. There'll always be an assortment of job related tasks to choose from as you go through your working day. It's an open invitation for procrastination. Moving seamlessly from one task to the next is a big problem for some techs. As stated before, there's no "later" in the world of IT. Later on you'll be just as busy as you are now. Learn to move quickly and directly to every new task and complete it with a "first time, every time" level of performance. Everything will start to get easier for you after that. Delaying the start of any particular job postpones the inevitable and only adds to your stress level down the road. It's far better to preempt the inevitable by aggressively meeting it head on. If another higher priority job pops up along the way, then move to that other job quickly and resolve it solidly so you can get back to your daily task list without missing too many steps.

Of course there'll occasionally be the big system that breaks down and becomes a major emergency that solidly interrupts your work flow, but realistically, most important jobs are actually just medium priority things that you can reasonably work into your schedule without undoing your plans for the day. Constantly dropping one job to go after another is a good way to lose track of your workload and make everything seem harder. Keeping a list of chores that you can check throughout the day is a good way to keep both your sanity and your work load on track.

- **Outside distractions**

Outside distractions can cause even the most vigorous tech to become a time waster. With the level of focus that IT work requires, outside distractions don't have to be great to have a negative impact. Even relatively minor distractions can break your focus and affect your job performance if they arrive at exactly the wrong time. It's not so much the magnitude of a distraction as it is the timing. That's why distractions are often described as "couldn't have happened at a worse time." Big or small, all distractions are the result of bad timing. A friend politely tapping you on the shoulder can be a major distraction if you're focused on something in front of you enough.

Distractions come from a wide variety of causes both inside and outside the workplace. It could be an obnoxious boss, bullying co-workers, a difficult customer, problematic vendors, an extended power outage that affects your home (but not your office), or never-ending bad luck with the family car. The thing to remember is that computers don't care about those distractions and how they might be affecting your work. Computers continue running while remaining blissfully ignorant of the human condition. What does this mean? It means that if you've chosen to work on computers for a living, then the professional part of you must adopt the same level of human detachment that the computers you're working on have. This may sound harsh, but do this or you will find yourself in the wrong career. It not so much that distractions are to be avoided, because they can't be…not in the real world anyway….it's that you need to develop the skills to manage distractions when they get in your way. Otherwise those distractions will only cause you to suffer stress.

This's where you have to decide your level of commitment to your chosen profession. You must decide what kind of IT professional you'll be: the kind who always stays on task or the kind that allows distractions knock them off track. Excellent techs are always in the former group. Balance in life is always important, but more critical is your acceptance that maintaining complicated machines that run twenty-four hours per day and are used by a lot of people requires more than an average amount of commitment on your part. You must have enough pride of ownership in your systems that it really matters to you how well they work around the clock.

- **Balance is everything**

When you begin with a new employer, the first thing to do is develop a good balance between work and home right away. Being the tech who always keeps their nose to the grindstone may bring you success at first, but that's only in the short term. It can also cause problems in other areas of your life which turn into distractions that slow you down in your place of employment. It's far better to think ahead and find a balance that accommodates both the needs of home and needs of work in an all around way that's as good for both as can be. Class A workaholics may look like heroes, but they are shooting stars that only shine brightly for a moment before burning out into oblivion. The idea is to find a way to have it both ways. Be the volunteering, hardworking type in the office while staying involved in life outside the office too. It's in how well you organize your daily work with your daily life that determines how sustainable your excellence will be over the length of your career.

A big step toward finding this balance is learning when to say no. Saying no is one of the hardest things for many people, especially young techs, to learn how to do. What's more, it also must be done with great deal of care to avoid the obvious negative repercussions. Nevertheless, it's a critical skill that'll allow you to move through your career in a manner that never compromises your standards or your good performance. For example, by saying no to a situation where there's a chance you'll do more harm than good, you'll keep stress at bay and your credibility intact. In fact, if more professionals said no to taking on jobs they weren't quite ready for, there would be fewer bugs in the world than there are today. Saying yes to working on a system you're unfamiliar with or to taking on a job when you don't have sufficient time to complete it correctly leads to mistakes which can make you look less than professional despite your good intentions. Your work will suffer no matter how courageous your efforts are. When your work suffers, so will your reputation. Saying no to being over extended is the first step toward taking ownership of your craft. You're a professional and you should earn your living by following professional standards. If you can't do a particular job well for whatever reason, it is better you don't do it at all. This isn't as hard to do as it may sound. Keep in mind that mistakes at work, in all cases, are always the result of a lack of balance in the workplace. No matter who gets blamed or how far you boil the source of the mistake down to its barest elements, it's always a failure of management in the end. It was either a failure of upper management for pushing a job agenda that was too unclear or too hard to follow, or it was a failure of a tech's personal time and/or skill-set management affecting their work.

Saying no to being over extended is the first step toward taking ownership of your craft. You're a professional and you should earn your living by following professional standards at all times. If you can't do a particular job well for whatever reason, then you should consider not doing it at all. Your boss is probably aware of this too from their own experience so let them know up front if you think there could be a problem with a task they're assigning you. Your honesty will keep them out of trouble too. After all, what happens in their managerial area affects how they look to the executives above them. If your boss is smart, the honesty will be appreciated and they may even offer the training or the time needed to let you still take on the task, but under better circumstances.

Balance and stress avoidance go hand in hand and both involve thinking ahead. No area in the field of IT allows for more preemptive thinking than does system design. A great way to find balance in your life is to approach a job from the outside-in. That is, the best way to avoid stress is to respond to the broadest requirements first to meet feasibility and then scale back to meet the details with increasing granularity. Always start with the complete scope and add details, starting with the largest parts of the project and then narrowing to smaller and smaller details as you proceed.

The opposite of that is trying to work randomly through the project's timeline. That causes a loss of perspective of the objectives and happens every time you try to work in a non-linear manner from a clear starting point to a not so clear endpoint. It's better to begin with clearly defined objectives from start to finish including all the milestones in between. Once you have a solid grasp on the overall scope of the project, follow by adding the finer details along the way while never losing perspective of each detail's impact on the overall job.

Once you begin, if you start to lose track of those well planned details you're setting yourself up for sleepless nights. This's another area where saying no comes in. Once you've started a job, do your best to avoid accommodating "helpful" comments from well minded bosses or teammates when they offer what they think is a better path for you to follow. What they're doing is deciding after the fact that they have a better idea than you about how you should proceed through the project milestones. Don't let them fool you. It's a trick…even if they don't know it. This sort of thing always leads to confusion, so anyone aspiring to have an excellent legacy in IT will plan ahead for these kinds of "helpful" distractions too. If you put together a clear job plan at the beginning, you can always wave it in their face when they decide to offer helpful changes later on. This's no small thing. Many projects have been knocked off track and even ruined by helpful meddling after the fact. Always maintain your original vision and be able to clearly define it in case it never needs defending.

You should never work open-ended on a job. Good planning ahead of time allows you to stand up and defend those plans once they're underway. Boundary setting keeps your job manageable from start to finish. While having the flexibility to change plans mid-stream might make you an effective military general, it'll always makes you a terrible computer technician. For IT professionals, changing plans mid-stream only means your original planning was badly done. You might be amazed at how many times your first idea turned out to be the right one after all. Changing plans mid-stream only muddles jobs, causes sleep loss, and, worst of all, usually ends up giving you a poorer result than what you originally intended.

- **Planning how things shouldn't go is as important as planning how they should go**

Planning how things can go wrong means bad ideas are never a waste of time. Even if an idea sounds like a terrible one, you should at least take the time to understand why. Otherwise you'll never know what failure states you might come across as you work through your task. Every good tech knows about setting boundaries for a job they're working on and part of those boundaries is hearing what may seem like bad ideas while the planning phase is underway. Any tech who just sneers at another tech's bad idea instead of taking into account the variables it represents will never be fully prepared when they start a job. If fact, they could be just an accident away from a hard lesson.

Don't be that kind of tech. Working through bad ideas before you start a job helps to avoid dealing with them at a later time when things can really go badly. This's how bad ideas can actually help reduce stress. The more bad ideas you work through and eliminate before starting off, the less bad ideas you'll stumble across later on. Don't discourage bad ideas in the beginning because they set the same kind of boundaries that good ideas do by forcing you to understand exactly what's bad about them. That means going back over the job in detail and knowing the true impact of every decision you make. Yes, it may occasionally seem frustrating when you're ready to start and another bonehead idea pops up for you to deal with. However, the more you do this at the start, the more stress free it'll be when you're underway. So always take time to understand why good ideas are actually good ideas and have enough detail of understanding to correctly refute the bad ones.

- **Build a silent solution**

A silent solution is one that runs quietly and does everything it's supposed to do without attracting unwarranted attention. The idea of being silent is to create solutions that present themselves like the proverbial iceberg. That is, most of the solution is out of view of the people using it. The less that's seen by the end-user, the less there is for them to complain about. Always remember that the typical end-user already has enough stress in their life from their own careers so don't need you adding to their load with poor system design or poor maintenance.

Of course, practically every IT system in the world already fits the bill of being mostly invisible to the end-user. From e-mail platforms, to databases, to intranets, and so on, most of the mechanics are outside of the end-user's field of view. The problem is that while this is how systems are supposed to exist in the workplace, too often they don't. There are overly noisy systems out there that expose flaws in their inner workings on a regular basis to the people using them. Marginal performance in those noisy systems creeps into the end-user's work space and bothers them throughout the day. Whether it takes the form of slowness or system hangs, unexpected error prompts, clumsy interfaces, etc., the system is no longer silent. It's now being very noisy. Problems caused by noisy back-ends are endemic to the entire company and give end-users rightful cause to distract you from your other tasks with persistent service calls to correct the IT issues they face. If end-users are continually troubled by systems that should be silent, then your sleep at night will be troubled too. Have enough pride of ownership in the things you build and maintain to care about making them quiet and effective at all times. Your end-users will demonstrate their appreciation to your efforts with their silence as well.

- **Learn how to learn quickly and continually, then remember what you need to know**

    Learning new things quickly is one of the most critical skills a tech can have. Being able to learn quickly is essential to a successful career in the IT industry. It's also the way to gain respect around the tech department. Even the most annoying tech bullies won't mess with someone who has good learning skills. Your good learning skills will leave them in the career dust every time. If you like respect and relative peace of mind in the workplace, learning new things is always the first place to start. Even if you run a network on your own, regularly increasing knowledge is the best way to resolve worry so you can work more stress free. Learning doesn't just come from books and how-to courses either.
    If you're surrounded by good techs, and you probably are, listening to others whose skills you trust is a   helpful resource too. Never be too proud to not take advice from others when your common sense tells you it's pretty good.
    This isn't to be confused with allowing others to disrupt a job that's already underway. Even good advice can be bad if it arrives at the wrong time. However, treat ideas from others whose skills you trust as one more resource when get-

ting up to speed on something new. Learning is part of the planning phase for any job and the more resources you can bring to bear, the better. The more you know up front, the less you'll be distracted by a lack of knowledge later on when the work is underway.

- **Maintain a sense of realism with everything you do in your work**

Never give an estimated time of repair you can't meet. Never make promises you don't have the resources to implement. Never give cost predictions that aren't accurate. Always stay within the bounds of the job in front of you. While it may seem so on occasion, no tech is actually wrestling with a system to keep it online. Computers are not a clever adversary you must constantly adapt to. Computer systems are static and, at least while they're well maintained, unchanged from the last time you or your department worked on them.

The only changes those computers will ever see are a direct result of your actions in one way or another. This includes negative factors too; like poorly managed LAN room temperatures, conflicting software patches, or even malware from an outside source infecting your systems. After all, those things wouldn't have happened without your help, whether you realize it or not. You're the caretaker of the operations or processes you've been assigned to develop or maintain and as such, any negative impacts to those operations or processes are ultimately your fault. This's still the case even if you didn't create the impact directly through your own actions. It doesn't take much for this to be true either. The fact remains that issues such as careless environmental management and weak perimeter defense are ultimately both products of your own efforts just by you allowing allow them to affect your area of responsibility.

To pick up on one of the examples, you can't blame a hacker for breaking into your system and causing problems. He may think he's brilliant, but he's just fooling himself. You're every bit as smart as he is; he just out worked you. He was able to disrupt your typical workday by trying harder to invade your systems than you were trying to defend them. Don't blame him for this. You systems being infected with malware is the result of your own action or lack thereof. Stay on top of the areas you're responsible for and they'll cause you less stress over time.

As long as you have a complete understanding of how a system works and interacts with other systems, it'll always be a relatively easy task to realistically estimate every job involving that system. Keeping your systems as static as possible ensures you'll have the power to plan for every contingency with great confidence whenever you work on them. However, even when you're working on systems you know well, it's still important to take a moment and think ahead, even when it's just a simple fix. Take a few minutes to review, and go over the processes in your mind before you begin each task. Preparing for a job by focusing on getting all the details realistically evaluated before you begin is the measure of how professional you are.

- **Remember you're the expert**

    When going about your workday, do your best to resist outside requests for some new feature or functionality that you know isn't right for the system involved. Consider this a call to arms for all IT professionals. You can achieve excellence by remembering that you're an expert and really do know more than anyone else about the systems you've designed, built, or maintain on a regular basis. They've been designed to work a certain way for a reason, so don't let any coworker, boss, or customer talk you into taking those systems to places they weren't designed to go. Use well written change control protocols and stick to them. It helps to include in the protocols a few preemptive points to better underline why something works the way it does. Doing this builds a barrier against misplaced notions that'll negatively affect operational continuity. If a big change is called for, then follow the IT department's approved system design life cycle approach, which should include the usual clearly defined user requirements, validation testing, quality assurance documentation, and the end-user's Service Level Agreement...like every good IT department should.

If it's a small request, then have an approval process in place that includes the requester's boss, along with plenty of time for testing before implementation. Do your absolute best to block the one-off end-user requests for something special on their computer or system. Even if it makes you look a bit snippety with the requester, never forget that the steepest slippery slope in IT is the end-user rationale that has them asking, *"They have it on their computer, so why can't I have on mine too?"* If you want to avoid stress, learn to see where even a seemingly harmless one-off request might lead to over time. If there's any chance if could lead to a piece-meal, on-demand implementation done haphazardly around the company, then stop right there and turn the request into an appropriately scaled project. Otherwise, you're sure to experience stress trying to accommodate an environment that'll become less organized with each new modification. Don't ever let yourself be lead around by the nose. Those are your systems, so always take the lead on everything that happens to them. How well you maintain your bearing when maintaining, and even defending, an appropriate sense of operational scale with the things you work on goes a long way toward defining the level of excellence you'll sustain throughout your IT career.

End-users

What's another big cause of stress? Simply put, it's your end-users. You can't blame them either. Remember, to them, what you do is voodoo. You have mystical powers over what happens in their office every day and as with anything mystical, myths and presumptions abound everywhere. What's more, myths may not even originate at work. They could be simple notions that people bring in from home or school. Myths come from articles read in magazines, heard on the radio, or seen on TV. What's more, computer myths can be propagated by anyone in the company, including executive management.

Every tech who has worked for any length of time has heard a myth or two. Myths can range from how to make older PCs run faster with a nifty piece of software, to how some customized system that worked at someone's last company will work great for this company too. The best way to deal with myths is to go back to the concept of the silent system. The less of a system your end-users see, the less they'll mythologize about how it can be made better. As long as their computer screen displays their work the same way it looked the day before and the printer makes the their data appear on paper without a hitch, they'll usually leave well enough alone.

A good policy that helps keep the systems silent is to never do favors for end-users. The list of favors is endless: install some favorite software they brought from home, downloaded a plug-in that provides an easier way to do things in their work-flow that's non-compliant to your department's design protocols, give them access to restricted things on the internet or a special functionality they don't really need, extend secured access just for a while, and on and on. There are dozens of examples out there, but don't do any of them.

There should be an established baseline for all personal computer systems used in the company. Any deviation from that baseline will increase administration as well as cost and be a potential source of stress. No matter how nicely the end user asked, any deviation from the baseline must be rejected. If you're a contractor, it should at least be reported to the IT department, even if it makes you look like a snitch. Non-standard requests will get you in trouble every time. Accommodating new user requirements in the production environment is a matter of protocol and the proper procedures for design and implementation must be followed. Later, after protocols are followed and a proper roll-out of new software or systems begins, there's nothing more frustrating than getting hung up by some end-user's personal solution that shouldn't have been running on their computer in the first place.

You must learn to say no without mercy, even to big bosses who are convinced that their personal stock market software reminds them that success is all about corporate value. Tell your end-users anything. You don't even have to lie. Good reasons for maintaining a strict baseline abound. If you tell the big boss that computer baselines save the company money, you'll be telling him/her the truth. Telling middle management that a strict baseline helps ensure uptime and a steady flow of billable minutes is also speaking the truth. So, when you turn to the end-user and say, "sorry, but it's the boss's orders," then that will be true too. Say whatever you want, but say it clearly and honestly. When you have your act together, what you say will be right. It's your profession and those are your systems, so don't let somebody rollover you with poorly understood requests. This is a key when it comes to stress management. Being on top of your job expectations is the best way to stay above any stressful situations the job may carry.

A lack of involvement by the end-user is always the best measure of how well things are running in your network. Even when you inevitably make changes to your computer environment, those changes should always be as quiet as possible. What's more, once they're using the new solution, they should never feel the need to think back to the good 'ole days when the old system seemed to work better. Your job is to make it possible for them to drop into the new environment as quickly and quietly as possible with a solution that's an improvement over the old. The best way to ensure end-user satisfaction during a time of change is to take the steps necessary for a seamless implementation of the new system.

A major part of systems design is to provide new ways for end-users to perform tasks in a manner which is an improvement over the old methods they used before. It's important to also ensure the end-users get good, user-friendly orientation on how to best work with the new solution. It may be a new release of a familiar software package or a whole new application for them to learn and use. Regardless, the quieter the new solution is implemented, the less stress it'll cause to you and your department. Nothing is better than happy end-users who like the new tools you've give them. Life is good when that kind of positive result happens. It's not a hit-or-miss result either. If end-users aren't happy and you're feeling stressed about it, don't blame anyone but yourself. If your boss is forcing a new system on everybody that he likes but hasn't thought through very well, it's your job to confront his bad idea with overwhelming poise and professionalism. Remember to keep things realistic. Use reasoning to get the systems back to where you know they should be.

If you ever find yourself rolling out a bad solution to make someone happy, you're not doing your job as well as you should. Stress aversion includes maintaining those skill-sets required to be able to say no to management on occasion. This's especially true when their expectations, as well as the end-user's expectations, are not founded enough on practical reality to be successful.

Overreaching Your Limits

Working on the edge leads to nervousness which can make anyone tentative on the job. A tech's nervousness gives others the impression that the tech may not know what they're doing, which of course only increases the stress you feel. Overreaching happens any time someone works on something they don't fully understand. You may find yourself in this situation on occasion. Sometimes no one in your department fully understands a particular system for whatever reason, so you've been "volunteered" for the task. Perhaps it's some legacy heap stuck in a back corner that needs one more troublesome patch a few months before its planned retirement from use. It could also be a system set up by contractors who didn't leave suitable documentation behind. If you want to avoid stress, take the time to get to know a system first before attempting to work on it. To pick up on the previous example, legacy systems are usually kept around for a good reason. Yes, it's probably sustaining some out of date software used by just a few people, but it may the only system those people have to work with for that one task. Screwing up a legacy system will not only make you look bad, but your department could look bad too. Nobody works well when they're unsure of their abilities and a situation like this can easily make you unsure. Just remember, the more you know about it, the less nervous you'll be when you start to work on it. The less nervous you are, the less stress you have to deal with.

Sometimes overreaching is caused by a well-intentioned professional filling in for someone else, even though he/she doesn't know the other tech's system all that well. Even a certifiable computer genius can lose credibility if they screw up a system they don't quite know how to work on. Excuses about how you were just trying to be helpful aren't allowed here. Always remember that competency is your responsibility. Moving forward without competency is, in all fairness, your fault and no one else's. When you're about to take over the maintenance of an unfamiliar system, you should stop, take a step back, and absorb all the learning materials required to be competent. Doing anything less is less than professional. If your boss can't wait and wants it fixed right now, then get those communication skills going and explain clearly to him/her why you should move with more due diligence. Being appropriately cautious will never cost you as much as making a mistake that affects company operations.

Worse still, systems only perform as well as they're understood by those who maintain them. A hurried solution can cause serious problems that aren't seen right away, but instead show up down the line…sometimes days or weeks down the line. If no one in the department is completely familiar with that system, then the latent negative aftereffects of your work can make for some doubly perplexing troubleshooting when they finally occur. Therefore, you must rise above demands for reckless expediency and make sure the work is done with the same professional approach you use on your regular systems in calmer times. Never compromise on this point because rushing in and making systems worse will damage your career. The result is a major source of stress.

Not Delegating Properly

Not delegating properly or letting too many people into the decision making process is another harbinger of stress. Too many decision makers will lead to too many paths toward a single solution which always leads to confusion. If you're in charge of a project, then the best way to handle having too many decision makers is preemptively. Set parameters at the beginning of the job, including the number of people to be involved in the decision making process. It's also good to clearly define what each person's role on the job is. Whatever the number of participants, be sure their roles won't overlap and cause design, implementation, or other conflicts that can sidetrack a job.

The main area of conflict with projects is usually ideas, followed closely by schedules, and then by levels of team commitment. Some people may simply be along for the ride. They show up in the beginning so they can claim some measure of success at the end without doing much in the middle. All of these issues will lead to stress if you let them. If you were handed a team by management, then lay down the law early and let everyone know the role you're assigning them.

Having too many decision makers on a job always leads to an unsatisfactory combination of compromises. Jobs should have only one objective and a single plan which clearly defines the best path to that objective for all participants to follow. Your vision for the job at hand requires a great deal of focus. Too many decision makers having a say in matters can diminish your focus and leave you with a foggy view of the project's endpoint. You should never start a job until you have clear vision of the path to the endpoint and the deliverable objective and what each team member's role will be along the way.

Allowing too many influences from too many people will doom you to high levels of stress. You must do the hard math and decide what's going to make you look worse: openly disregarding all those helpful decisions made by others, or failing to complete the job successfully and on time. A good rule of stress management is to know when helpful decisions start getting out of hand and where to draw the line. Be firm but not rude. Remember, they're still your teammates.

If decisions are being handed to you by people you can't disregard, such as upper management, then figure out as early as you can how to take control of them. The best tactic is to have clear guidelines for all decision makers before you begin a job. If the guidelines are well developed into the project plan, then even upper management will abide by them. Create guidelines that provide guidance in the same way a road detour sign moves traffic around a construction site...meaning your project plan should include guidelines that'll steer anyone who strays from the plan back onto the path you originally set. You can do this via such things as a well drafted Service Level Agreement approved by all parties or a detailed project timeline that's too well designed to be modified at a later stage. Even signed vendor agreements which can't be reneged for a different option are tool you can use to lock down your planning stages. The idea is to deflect meddlesome decision making with preemptive action that defends the path you've laid out for the job. The better laid your plans are at the start, the less opportunity others will have to change them once things get going.

The same goes for simple jobs such as a service call, although the notion of too many decision makers here is a slightly different concept. When it comes to doing an on-site repair, even two people can be too many. Yes, helpful advice can occasionally come from back-seat drivers, but usually their over-attentiveness is a distraction that can hamper your ability to stay on track. The problem in the IT world is that you don't ever want to alienate those around you because you'll always benefit from their help and advice over time. To keep the work environment positive, it's best to develop skills to avoid having unnecessary advice offered to you in the first place.

Best practice in this situation is to explain your actions as you go. If your coworker(s) know what's ahead, they'll be more inclined to go along with your choices once you move to the next step. It may feel like talking to a child when you explain each step you're taking, but it won't be heard that way. It'll sound to them like you're engaging in appropriate discourse as you move through the job. Once you begin working you'll be able to proceed from step to step without the stress of someone questioning your every decision at each turn. They won't need to because you've already told them what those decisions would be ahead of time.

If they insist on offering advice after you have started the job, then you need to reach down and do a gut-check of both your pride and your confidence. First, check with your pride to make sure you're not just turning down good advice on principle...i.e. you don't like someone else knowing more about what you're doing than you do. This kinds of stubbornness can lead to embarrassment if you eventually do the wrongs thing. While ignoring good advice is seldom a good idea, in this situation, you should take in account the source as well. Always keep in mind that even the best advice can be of no use if it's delivered poorly and there are few poorer ways to deliver advice than waiting until someone has already started to work on something before speaking up. If the advice giver was truly conscientious, they would have piped up before you began, especially if you're letting them know your intentions ahead of time. On the other hand, maybe they're just being annoying...it's your call.

Once you've checked your pride to make sure you're not just ignoring a better solution because it wasn't *your* solution, then once again it's time to check your gut. If you know the job you're doing is the best one for the task at hand, then trust yourself and ignore the other person. Don't be rude about it, but you do need to be firm. You should explain to them that this's the way you want to proceed and you're confident it's the right way. Even if you make a mistake at this point they'll at least be more sympathetic. Plus it'll help you hone the leadership skills your career will demand as your success grows in your field.

Whether you're working in a large team or with just one other person, a sure cause of stress is weak leadership skills. This's true whether you're the leader or a follower. Weak leadership is demonstrated by an inability to do things firmly, therefore causing a gap in confidence that'll affect the job. Remember from earlier in this book that while it takes courage to start the job, it takes confidence to keep it going. Anything that hampers confidence will slow the job down and cause stress at some level.

If you're just a team member on a particular job, you might have no say if it becomes disorganized by weak leadership. The problem is that if you try to tell the team leader the project is getting off track, you'll only be one more opinion they have to suffer through. If you're in this situation, step number one is to first confirm that you're not just being a whiner. This's important. Even the slightest hint of a whine in your voice could render your opinions mute to a leader with weak delegation skills. After ensuring that you're not whining, the next thing to do is to back off and keep your mouth shut anyway. This isn't meant to be trite; it means you should take the time to find a better medium of communicating than just adding to all the noise the leader's probably already dealing with, regardless of how valid your points may be.

The better way to communicate is to take a higher road that your coworkers by delivering your opinion in a more professional manner. Put your points into a brief but accurately written statement. The statement can range from an easy to digest email to a comprehensive report; just make sure it's appropriately scaled to the work the team is doing. The team leader will appreciate this because he/she can review your comments at a later time in a less hectic environment. This also ensures that your opinion will be read uninterrupted, so it'll carry more weight. Just one good, clear, pertinent statement on a page or two is all that's required. Even if the team leader rejects your suggestion, they'll at least appreciate the time and effort you put into writing it. In effect, you're helping the troubled leader with their stress management too.

Delegating Without Checking First

Another form of poor delegating is relying too heavily on others who are not as capable as they should be. You should always judge delegation carefully. It can be a risk to your credibility if the person you delegated a job task to ever fails you. If you made the call that allowed others to drag you down, then you've demonstrated bad decision making skills. How you can get into these predicaments varies with each situation, but there are a few common causes to consider.

If you are trying to mentor a younger tech by giving them a tough task, then gauge your role carefully. Mentors are important to IT because creating a learning opportunity to beginners will help grow the quality of the IT professionals around you. In fact, in an industry that reinvents itself every three years or so, mentoring is critical. Just keep in mind that delegating tasks to others who are less qualified than you in hopes of providing some needed experience can be a dangerous thing to do. Even if nothing comes of it, the danger will still add a bit of stress. Make sure you monitor as well as mentor, so that any actions taken by the less experienced tech can be amended if they ever head in a wrong direction.

The best means of mentoring is to use a purpose-built bench environment. All IT departments should have some learning environment for techs to use when training on new systems. This could be hardware left over from a former development project or something permanent that occasionally doubles for the testing and validation of systems in production. It could also be a pure learning environment where new skills can be developed on systems that aren't attached to anything at all. The more savvy IT shops usually set aside some equipment for this purpose in a side room where techs can try things out during downtime.

Since stress comes from allowing others to make you worry, be careful when delegating tasks. Someone who seems qualified and someone who *is* qualified are two different things. Always have a complete understanding of the depth of skills the people around you have so that you can distribute responsibilities knowingly and carefully. If you simply point to the first person in line and ask him/her for help, you're looking for trouble. On the other hand, if you have too low of an opinion of your team members to be able delegate anything at all, you'll end up doing everything yourself. It's critical to remember that both outcomes are the result of the same thing: allowing the skills of others to lead you into a stressful situation that could affect your credibility.

You can become knowledgeable of other tech's skill levels by simply communicating. If you have doubts about how well another teammate can handle a job that also affects you, just ask them in a kind manner that lets them know you only want to be stress free about the whole thing. Mutual reassurance between team members who work together in ongoing projects is important in any business. It's especially important in IT where it's impossible for everyone to know everything. It's best to learn to talk to others about who's the best choice to take over different types of tasks.

Forgetting to Delegate in the First Place

The other side of the coin is not delegating at all and trying to do too much on your own. Techs working in a one-person shop probably understand this better than most. They have no choice but to do almost everything on their own. However, team oriented IT pros have to remain connected to the others around them, even if they've been burned by bad help in the past. Team techs must never lose trust in their teammates. Trying to do too much on your own may seem self-gratifying for the moment but quickly leads to stress for obvious reasons. Taking on more than your fair share will leave you exhausted which will not only result in stress but, more importantly, in mistakes too. If those mistakes cause problems for the end-users as well, then the stress becomes contagious. Pretty soon everybody in the company is feeling stressed because of your lack of trust in your co-workers.

Don't ever believe in the antidote, "If you want something done right, do it yourself." The guy who dreamed up that nonsense never knew what it felt like to be truly busy. If the workload can be shared, then it's important to share it. If you don't have confidence in someone else's ability, then it's part of your role as a teammate to try and come to terms with that. Picking up a struggling tech's workload is not heroic, it's a bad habit. Not only will the struggling tech's skill-sets not improve through use, but you'll also be working beyond the point of a reasonable work load. Bad habits like this can affect your career. An excellent tech will find a way to make sure all other teammates are part of the overall effort and are carrying their share of the load.

In some cases, such as an upcoming hospitalization, selected service duty, or some other planned leave, coworkers may voluntarily take on an extra job or two in support of the individual who is not available for work. This's noble, but can only go on for so long. It can also add to the noble volunteer's stress level so must be monitored carefully to prevent becoming a problem for everybody.

Of course, the one person shop doesn't have the option of sharing the load. Everything comes down to him/her. Even if the company has only twenty employees, if those employees are busy all the time, the tech will be busy all the time too. If you're a lone-wolf tech who's feeling overworked and asking your boss to hire a second tech to share the load isn't producing results, then being well organized and preemptive in your tasks is the way to go. If you can see what's coming down the road, you can react before it reaches you. For instance, studying some technical guides on a new system your company's talking about but hasn't signed off on yet is a good way to stay ahead of the curve. If/when they go that direction you'll be ready to go ahead of time. Of course, this approach will occasionally lead to the uncomfortable situation where your preemptive work amounts to nothing because your company decided to go in another direction after all, but don't let that frustrate you. Even if your effort on that particular occasion was wasted, you were right to put forth the effort anyway. Had they moved in that original direction, you would have been one step ahead. So they moved in a different direction; it's your job to see that coming too and have already moved in that new direction as well of them. This's the plight of the lone-wolf tech running a one person network; you must always stay one step ahead or the stress will quickly eat you up.

The same goes when maintaining the systems you built into production. The more preemptive your maintenance is, the smoother your operation will run. Preventative maintenance is an important task in any environment, but there's no place where it's more critical than with the lone-wolf tech. If you are running the department on your own, you still need to keep up with industry advancements.

This means no holding on to yesterday's technology just to make your life easier. When the legacy stuff starts breaking, you'll truly feel alone when vendor support for those old systems dries up. It's always better to avoid the unsupportable legacy stuff as much as possible. That means pushing your boss a bit harder for an upgrade from that "oldie-but-goodie" system they've been using forever. In the long run, not only will the company profit from a more modern solution, your job will be easier too. The more supportable your systems are, the less sleep you'll lose at night.

Lack of Motivation

Nothing saps a person's motivation more than stress and worry. Health issues and personal happiness aside, the biggest impact stress has on you professionally is that it causes your motivation to lag. Lack of motivation is often the result of a chronically negative situation left unattended by either you or your company. Because stress can cause a lack of motivation and, conversely, lack of motivation eventually leads to stress, it's important to see this as a vicious circle. Each reinforces the other's negative impact on your wellbeing. To break free from this feedback effect, begin by gaining a better understanding of your motivation issues in general.

Lack of motivation can be broken down into two types: internal and external. An internal lack of motivation is what you see in lazy coworkers who have a low level of personal drive by nature. Their default path to being stress free is to do as little work as possible. Lazy techs are easy to recognize because they're the ones questioning why things need to be done to spec when doing a quick fix would be good enough. After all, the less they do the less they'll be responsible for. The problem here is that the less one tech does, the more everyone around them will have to do to compensate for their lack of output. This's why the IT professional with internal motivation issues will likely be a source for external motivation issues in the other IT professionals around them. How to deal with lazy techs is covered later in this book, but suffice it to say they risk being a liability no one needs around.

External motivation issues come from how the working environment affects a person's desire to work at a high level of output. No matter how calm and collected some techs appear when others around them are frantic, stress is always a departmental issue and not an individual one. No one is immune. Every busy workplace has a little bit of stress as a matter of course, but in a well managed workplace the stress is divvied out evenly to all members. It's the responsibility of management to ensure this's happening. A lazy coworker will simply takes less than their fair share of the overall stress, which causes others in the department to take more. In the end, the lazy person is simply avoiding stress by deferring it to others. Unless this can be cured though discussion, a lazy coworker should either have this behavior modified through punitive steps or, in a worst case scenario, be removed from the staff as soon as possible. Their presence can be poisonous to an IT department's morale.

External motivation issues are also apparent when a person has trouble speaking up. If you know something is wrong, don't wait for everyone else to find out about it after the fact. It may seem a good way to avoid the stress that accompanies the act of giving others bad news, but that kind of avoidance will come around and haunt you later on. If anyone finds out that you saw a potential problem but didn't speak up in time to prevent it, your credibility as a professional will be damaged. Even if you feel too unqualified to express a view which is different from what the more knowledgeable techs around you are saying, do so anyway.

It'll help you learn more about the system you're working on either by reinforcing that your reasoning is solid or by providing an opportunity for someone who really does know the system better to explain their thinking more clearly. You, and probably others around you too, will have a better understanding of the logic behind the job and how it's being applied. This helps you become more knowledgeable in your own right. Computers respond best to experienced handlers and experience often comes from the involvement you gain by speaking up.

Finally...

Even if you have your own personal stress solidly under control, if you aren't paying attention you may be leaving a trail of stress behind for other techs to endure. This happens when you don't document what you're doing well enough. Never pass on something you've to done to fellow IT professionals blindly. Always leave behind documentation that others can use to follow in your footsteps. The ultimate silent system is one where you never 'sign' your work by adding elements that will remind others you were there. Other than passing on a set of good documentation, have the professional bearing and strength to leave no proof the work was ever done by you. Besides, in the IT world, there'll be plenty of new jobs ahead of you to validate your growing talent. It's about what you can do today, not what you did yesterday. Old work is best left anonymous to anyone who hasn't read your resume. There's strength and good professional bearing in that.

CHAPTER FOUR

Understanding the End-user

The bond between you and the computer systems you work on is an essential part of your ability to manage complex tasks under pressure. You should never allow others to affect that bond. End-users need to be listened to and you should pay close attention to the costs that management want you to maintain, but no one is more qualified than you when it comes to determining how well your systems fit the company's needs. After all, that's the reason you were hired in the first place. Your job is to assist in building and maintaining the digital infrastructure that your company uses to do its daily business. Self-honestly about your limitations is a big part of this, but so is your belief in your own judgment regardless of how critical others around you may be. How well you deal with end-users will speak loudly to the level of excellence you can achieve in your IT career.

Confidence when working with others is critical. You have a clearer view of the systems you build and maintain than anyone around you. If you know a system is running properly and meets the corporate requirements which guided its design and implementation, then criticism from end-users is not so hard to take. They usually don't understand the true depth of the computer tools they're working with anyway. It's difficult for the end-user to understand that the front-end application you provide them is really the result of many different back-end applications patched together just right. All they see is the front-end on their desktop. If that small part of your solution is criticized by end-users for not being the best as they see it, keep in mind they don't have access to the overall scope. The concept of the whole exceeding the sum of its parts is very apt in the world of computers. You know how all those parts fit together on both the front-end and on the back-end in ways that the end-user never sees. If some aspect of a system may seem weak in a certain light, it might not mean there's a problem. It's the overall effectiveness of the complicated system that will be judged in the end.

However, even the most balanced systems can be fleeting. As is typical in the Information Technology industry, a perfectly effective solution will only be around for a few years before it's obsolete. For your career to be self-sustaining, you must measure your understanding of how well a system meets the end-user's requirements with how well you can deal with those inevitable end-users who complain even when you get everything right. If people who know less about computers than you do are able to hurt your confidence through complaints about a small part of something you've built, it's possible you'll never design a good system again.

Listening too much to criticism will make you over sensitive to it and that can lead to major problems down the road. Oversensitivity to criticism causes you to forget about things like balancing user requirements with cold hard cost benefits or properly measuring procurement schedules against project time limits. It takes a lot of skilled patchwork to put complicated systems together in a constantly changing world. Solutions should not be judged up close and piecemeal, but at a distance that's more appropriately scaled to the company's overall needs. It's commonly thought that striving toward perfection is the hallmark of a successful career. However, in the rapidly changing world of computers, the concept of perfection needs to be carefully melded with real world practicalities, such as budget constraints, user orientation, and anticipated obsolescence.

Never lose track of the overall picture. Don't let any criticism of one part of a system affect your confidence in the system as a whole. Trying to make everybody happy might lead to level of acquiescence on your part that could, and often does, have disastrous results for the systems you're working on. You'll find yourself keeping the old stuff around too long or integrating newfangled technology before its time. You'll lose track of how badly system obsolescence can collide with new solutions coming on the market. You'll be putting together solutions to keep everyone happy despite knowing they're not the best solution for the company in the long run. If you head down that path you'll end up with user requirements that'll be all over the place and find yourself engineering conflicting needs into the system designs, spelling doom from the start. If you've already gone down that road, it's time to take a deep breath and get your confidence back.

Start by listening to the computers themselves before you listen to the end-users. Get the systems you're putting together to work well all the time for everybody and the end-users will come around. Why? Because they have better things to do than bother you. End-user rankle is almost always temporary. If their complaints have true merit then of course you should listen and respond quickly with a better solution. However, if your systems are working in a reasonable manner then it's likely the end-user complaints will not be universal.

If you look around you when you're in that situation, you'll often find that for every end-user who has a complaint, there are many others in the office who are working with the exact same tools just fine. If you can see that the systems are running as designed, then use that knowledge to communicate with the end-users with confidence. You must stick to your guns and fight though any obstacles a complaining end-user throws at your feet by displaying the wisdom and expertise that you as the computer professional bring to the job. If you are honest with yourself about how good the systems really are, your credibility will carry you through a struggle with a difficult end-user every time.

That being said, it's important to stay in close touch with exactly what the end-user is seeing on their monitor each day. To measure how good your systems actually are, start by looking at things from the end-user's point of view. For example, a typical office worker starts their day with a cup of coffee and a quick breakfast before begrudgingly setting off on the long morning commute to the office. The commute is never fun, plain and simple. It's a daily struggle that takes too much energy and leaves people fatigued by the time they finally reach their desk. However, was that long commute a waste of time? Not necessarily. The end-user probably spends part of their morning commute mentally reviewing the upcoming events for the day. Maybe it's an early meeting, the report that's due, the inbox item that has a short deadline, a job order they got the night before just as they left for home, or any other important item waiting for them when they arrive to the office. They use the commute to review what they want to do as soon as they get to the work and then prioritize those things in an order that gets the most done as quickly as possible. It's how they intend, like every good worker, to get through the morning as efficiently and stress free as possible. This's also where you as a computer technician should never come into the picture.

If they spend even one measly second during the commute wondering if the company e-mail will be working, if their printer will be OK, if an application they need is going to freeze again, or if they have enough time to go through that really slow boot up to still be able to get their notes together before the first meeting, then you as an IT professional aren't doing your job right. It's that simple. The tools your company provides to its employees should always be functioning at the highest possible level or everybody in the company suffers. Those tools should never cause work interruptions...otherwise they're not very good tools. It's your job to make sure your fellow employees only have to worry about the daily tasks the company gives them, and not about the tools they use to complete those tasks. That's why if they spend even one second worrying about how well those tools are going to work that day, you're failing at your job. It may sound like a mean thing to say, but if you don't like the cold hard truth, that's too bad. Since your computers aren't ever going to let you off the hook for sloppy workmanship, and they aren't, why should anyone else?

Nothing should be less important to the end-user than the computer on their desk and whatever it connects to. It should be a button they push, some absent minded keyboard strokes, a few mouse clicks, or anything else they can do blindfolded. It shouldn't be anything more than that. They should never give their computer's performance a second thought. The more mystical it remains to them, the better. It's only there's to help them do the job they were trained to do.

When responding to user requirements for new or upgraded system, ask yourself how quickly the end-users will forget all about those new improvements and changes as soon as they've been implemented and how quickly they can get back to blind keystrokes and mindless mouse-clicks they were doing before. A true measure of success is how quickly something new in a company's computer environment can be taken for granted by everyone using it. The faster a tech can make this happen, the better they'll be at building and maintaining computer systems. Being in touch with the needs of fellow employees and the work they do while using one of the computers the IT department provided them is a critical part of a tech's bearing as an professional.

That's because the typical end-user is both a both a computer literate dummy and as well as a highly trained professional. They have enough of their own work to do without being a test case for how well your computer systems are running on any given day, that's why you were hired. The more ignorant they can be about that magical contraption on their desk, the better. They less IT mishaps they have to deal during their workday, the better you are at your job. For the company to be productive, your fellow employees must be able to enjoy the freedom of being dumb as a brick about their computer while still successfully accomplishing every task they're assigned on time.

Keep in mind that no amount of back slapping and cheers of applause from your boss and end-users will make this level of office functionality possible. In fact, back-slapping and heroics can be seen as a failure of sorts. A hard but very practical reality is that bosses and end-users have more important things to do than watch the IT department successfully overcome shortcomings in the computer systems that shouldn't have been there the first place. A tech might look like a hero running around putting out all those brush fires and saving the day each time something breaks, but what they're really doing is a marginal job. Every IT system their end-users work with that sits behind their keyboards and monitors should remain invisible to them and so should the IT department's staff.

A tech doing his job right is like the stage hand the audience never sees. A stage hand running onto the stage to push an actor out of the way of a falling spotlight may look like a hero, but really shouldn't have let the spotlight fall in the first place. From then on the actors will be more concerned about what's above their heads than they are about the audience in front of them. Their focus will be misdirected and their performance will suffer. The same concept holds true for computers, too. Anyone will lose focus on the job at hand if falling spotlights, or marginally operating computers, are constantly distracting them. It's better if all the behind the scene activities remain behind the scene at all times.

 If you can't manage to not be seen doing your job by your fellow employees, at least strive to be ignored. This may be a tough concept to grasp at first, but the sooner you make yourself look obsolete, the better you are at building and maintaining the silent system. That is to say, the end-user should never know the back-end exists. They only need to turn on their computer every morning and find what they need on their desktop the same as yesterday. End-users don't need to know any more than that.

Even if you're running around like crazy keeping the systems running smoothly, your fellow employees should still never know you were there. You should strive to be invisible, or at least underappreciated. Being underappreciated is a tough concept for many techs to take but it's an essential measurement of professionalism. The best networks are the ones that make everything look easy no matter how complicated they really are. Don't be afraid to be underappreciated. It's how you can best judge your accomplishments. In fact, a truly silent system should even make you appear a bit unnecessary. If the big boss is wondering why you're still on staff because the computers seem to be running fine all by themselves, then you've officially achieved excellence in IT. You're still on the staff because that level of silence takes an awful lot of hard work the end-users, and perhaps the big boss too, will never see but will always notice the moment you stop the hard work you're doing.

The thing about job security and job appreciation that's difficult for some IT pros to understand is that the success/accolade process is backwards in the field of IT. If you're going a great job, then a lack of appreciation from computer illiterates is a *good* thing. Everyone likes a pat on the back now and then, but being a superhero and being an excellent IT professionals are opposite concepts. Superheroes get attention and excellent IT techs don't. That's because while superheroes are running around saving the day, the excellent IT pro makes sure the day won't need saving to begin with. The excellent tech always stays ahead of the curve through good preparation and job maintenance so they can preempt problems before they arise. The best compliment you should ever hope for is someone giving you a passing comment on how there never seems to be any problems with the computers. If you never receive a compliment, better still. Only the IT department and the computers themselves need to know how good you truly are.

A great example of how the concept of a lack of appreciation works was the big moment in the history of IT when the whole industry, tech by tech, handled what was popularly known back then as the Y2K bug. Younger techs may need to be told that the Y2K bug involved old system clocks struggling with the "00" in the year 2000. Yes, the fact the bug existed in the first place was the product of short-sightedness and stupidity by an awful lot of smart computer people who *weren't* demonstrating excellence. Shortsightedness and stupidity seem to be at the forefront of all bad engineering, IT or otherwise. There were concerns the world's IT infrastructure would collapse as computers systems around the planet fell offline because of the clocking issues. As the year 2000 grew closer, some fanatics were even proclaiming a coming global apocalypse when the bug hit at midnight on January 1st. Nevertheless, while this could have created the kind of brush fire that makes for major super-hero status, it also allowed for a relatively quiet fix by a world full of technicians more interested in good computer operations than acquiring acclaim.

When January 1, 2000 came and went and the world was still standing, the IT industry received the best compliment possible when the world's news outlets collectively exclaimed "So what was the big deal anyway?" There were even a few conspiracy theories claiming the whole thing was just a trick to boost IT funding. Despite allowing a crappy environment to exist in the first place, a professional effort by the IT world got everything back on track as seamlessly and quietly as possible. In the IT industry, the unappreciated fix is always the highest form of professionalism. If you ever hear another tech complain, "They don't have any idea how hard we're working," you should remind them that they never should either.

The lesson to take from all this is that if your end-users are giving you kudos for filling in proverbial potholes that shouldn't have been there in the first place, then you need to review your performance. Eventually your boss will figure this out too, even if he/she doesn't realize it yet. It's the marginal systems themselves that will rat you out. If you build and/or maintain a system that doesn't work well, then that system just might keep you awake at night for a good reason: it's getting ready to expose its performance problems to the company. You're much better served by keeping the system quiet through first rate design and maintenance from the start, even if that means no one ever has a chance to notice how good you are.

The silent system is any system that runs smoothly all the time. This concept scales all the way up to the WAN level too. No matter how big or small your network is, there's never a good time for computer interruptions. Remember that computers are half of a human-machine working relationship. If the computers are interrupted, so are the end-users. Anytime an end-user struggles through an interruption, they lose billable working minutes and cost your company money.

Staying Ahead of Problems

It's no secret there will be the occasional individual who unfairly blames their computer for their work being late when their computer had nothing to do with it. You can't do much in that case other than defend your systems in an appropriate but truthful manner. However, the reality is that the vast majority of end-users are hardworking people, and computer problems really do break their valuable workflow. As a professional, you should see those problems coming and deal with them preemptively before they ever reach your end-users.

Using the same systems that your end-users use on a day to day basis is a good start. This gives you immediate feedback on how even the most simple tasks are performing on the computers used around the company. Some IT shops even make a point of issuing old but not quite broken computers from inventory to their staff to help keep the operational picture as clear as can be. Provisioning the best PCs and laptops to the IT department only blinds techs to how bad things may be out on the floor where older equipment is still in use. If anybody should be seeing glitches in applications used by everyone in the company on a regular basis, it should be the IT personnel responsible for those applications...not the end-users. Glitches always seem to be more apparent on older hardware rather the newer issue equipment.

Another helpful way to keep track of how well the computer systems are running for the employees is to perform some type of quick check of the systems at the start of every work day. These checks could be whatever is appropriate to your systems and scale from a simple visual inspection to real-time WAN level automated diagnostics with alerts that run around the clock. Regardless of how you monitor system operations, the point is you should never rely on your end-users to tell you that something is wrong. The systems should remain silent to them no matter how many entries may be buried in the error logs.

If a serious problem does pop up requiring downtime to fix, do your best to keep the repair as silent as possible. If that means waiting to work on the problem after everyone else in the company has gone home for the night, so be it. Chalk it up the twenty-four hour life of a computer professional. It's the level of commitment you apply at this stage that'll make your life easier over time. If everyone you work with is happy with their computers, you'll be happy too.

Constant Complainers

Even the most diligent techs must occasionally deal with an end-user who complains constantly when nothing is really wrong. Eventually, about all you can do is grin and bear it while surreptitiously ignoring most of what they say. If they continue to the point of agitation, just keep in mind that this's probably not an isolated experience with this person. It's likely you're not the only one who notices this tendency in their behavior. Their other co-workers may have received an earful a time or two on issues ranging from company benefits, to office seating arrangements, to snack room cleanliness, to middle management's competence, etc. If you're troubled by their behavior, then it's very likely the people who work with them every day are troubled too.

There's an old adage that "the squeaky wheel gets the grease." However, that expression only tells half the story. The truth is the squeaky wheel only gets the grease as long as people are willing to keep applying grease. Eventually, all squeaky wheels just get replaced. Don't let squeaky wheels get to you, and more importantly, remember that as a tech it's not your place to get involved. Let Human Resources or upper management deal with whatever morale issues that type of employee causes. As a tech, it's better if you just focus on doing your best and keeping the computers in the office up and running fine. Anything less and your confidence is at risk.

Going too far to accommodate the squeaky wheel causes two negative side effects you'll have to deal with later on. First, they could end up with a customized, one-off set up that must be reverse engineered every time their computer is upgraded. That means maintaining separate documentation just for them. Second, you set yourself up for a fall down a slippery slope when others sitting nearby become envious of the perks you're handing out to the complainer. The others may end up wanting some of those perks on their computer too. Tech's who accommodate one-off user requests always end up with too many computers running outside the accepted company baseline to be easily managed. Without the organization of a well maintained operational baseline, even the smallest trouble tickets could require extra review time and effort to resolve. It's best to say no from the start, no matter how much the squeaky wheel complains.

The insensitive approach applies to upper management, too. It's reasonable to suggest that every experienced tech has encountered a higher-up complaining about a system that everyone else in the company likes. The managerial complainer may even try to make their personal struggles with an otherwise operational solution appear universal to the whole company even when it's just affecting them. It's best when dealing with management to develop a few methods to preempt this kind of thinking.

A common way for dealing with a boss who has a habit of complaining about systems that everyone else in the company likes is to isolate them early on in an application's lifecycle. That's done by placing them at the edge of the roll-out schedule when the solution is first being deployed. Either give them the new solution after everyone else has it, or give it to them before anyone else has received it yet. In the first case, you can tell the boss the employees are already using it and are happy so the boss's concerns are unwarranted. In the second case it allows you to bring the boss into the pilot phase to get their "help" with confirming the solution is working "just right" before it's given to the employees. In either case, you get to keep your professional demeanor (even if you wish this kind of 'human engineering' wasn't necessary).

If the constant complainer is a customer who has access to your systems, then the issue is much more sensitive. Customer complaints can become political very quickly so make sure you've done your homework and can establish beyond a shadow of doubt that the systems are working exactly as they should be. It may only be an orientation issue or perhaps something outside of your scope of operations rather than a design or maintenance problem. Regardless, never respond to a complaining customer until you can back up your response with accurate statements. In this situation, don't be afraid to dive into the problem head first from the customer's perspective. That's the quickest way to see what they're seeing and gain a true understanding of where their complaints are coming from.

That may mean working directly with the customer's IT department to get the problem resolved. That's not a problem, but just make sure to only work with them after the inter-company communication is covered by a non-disclosure agreement. Also, be sure to share the load with the customer's IT folks to get the problem truly resolved. Don't let it become a verbal tennis game where you just keep hitting the ball back and forth into each other's court. It's a common nature to want to pass the buck to the other guy when the other guy's systems are outside of your purview. However, working together as a temporary team to resolve the problems make both party's jobs easier. If it becomes apparent after all the team efforts that the customer simply likes to complain, then turn the whole issue over the people in your company who work with that customer. Just make sure they have good access to your facts and that your communications are clear.

If the aforementioned end-user management tactics sound like playing head games, remember that people management is part of practically every corporate job role, high or low. Think of the human component as one more aspect of a system's design and maintenance schedule. After all, end-users are half the computer environment. In virtually all cases, computers wouldn't exist without end-users. Your job is to ensure that the human-machine integration is implemented and maintained successfully. If you want to refer to the 'people maintenance' part of that equation as head games, so be it. The bottom line is that the systems get moved to production and maintained as seamlessly and successfully as possible. That also means finding ways to work with those who require a case be made for their eventual inclusion whatever new solution is being brought online. Remember that you're not the only one who wants to limit stress in the workplace...your fellow employees want to limit stress too. Having to learn how to use a new piece of software or new steps in a new business process after being used to the old way of doing things is always a cause of stress.

Rather than accept the new change to their work environment gracefully, some people can resist your efforts every step of the way. They could become real roadblocks to the solution you're trying to implement. If you don't deal with them successfully up front, then they'll be the first to complain if any bug shows up in the system later on. Especially if a new application is significantly different from what they were using before to complete the same task. Whether you like it or not, this's just one more thing to add to your daily list of IT duties. Taking the time to communicate well with people who feel stressed by something you're bringing into their work environment will demonstrate your good bearing. Understanding the complete job, and your responsibility to accommodate more than just the mechanical computer half of it, will make you both an easier person to work with and a better employee overall.

Nonetheless, if end-user hindrances become so great that you risk losing focus on a job that's underway, then this's where the proverbial rubber hits the road with your confidence as an IT professional. When a tech starts to give into particular end-user demands that they know are counterintuitive to the project schedule and ultimately to company productivity, then they often find themselves laying awake at night thinking about work instead of sleeping calmly. If a computer professional ever suffers serious doubt when things start becoming too complicated by resistant end-users, then it's OK to turn away from the human half of the equation and back to the computers services being provided. Simply stay on the path you've set and remember it's the path already approved by company management. Ultimately, the rest will take care of itself.

Work as best as you can with any problematic individuals by maintaining a sense of empathy for their concerns and also through reassuring them that any changes being made were well thought out and are ultimately beneficial for the company. However, never let them disrupt your well laid plans. If you need to develop a more compelling explanation for them, the take the extra effort to do so; within reason of course. You can't spend too much time on this since the work still needs to get done. It helps to remember that if the two main measures of success are uptime and accessibility, then the only measure of a job well-done is ultimately the computer systems themselves. If you know the implementation was properly done and is well maintained, then your confidence in those systems will help you move forward. Eventually, even the most troublesome end-users will come around too once the dust has settled and the change become the new norm.

Knowing Your Systems

You must know your systems extremely well. Computer interruptions come at the worst possible time for techs too. It's not just the end-users who are inconvenienced; the IT Department suffers inconvenience just as badly, as well as losing a bit of credibility each time systems go offline. Interruptions cause a measurable decrease in workplace output, which means money lost from the company's bottom line. If the affected end-users are in a workflow oriented environment, their downtime will affect other business processes in the company. In fact, the economics of a disruption can reach further into the company's bottom line than the inconvenience of the original downtime did. This puts the IT department in a sensitive position when issues continually crop up that affect operational stability.

To know your systems well you must spend a lot of time with them. The more preemptive you are with maintenance, the better off the systems will be in the long run. This means not ignoring small issues when they don't seem to be doing any particular harm at the moment. Those small issues are only waiting around to make trouble later by adding their weight to the new problems that will assuredly come along. No fix should start by first working around small legacy issues that were left unattended in the past. Small issues can become big issues by slowing you down when trying to move rapidly through a critical repair.

Seamless integration should always be a component of every activity you undertake. It's part of being silent and unseen. If this seems impossible to achieve in your computer environment, it would really behoove you to stop at some point and try to understand why. There has to be a deep underlying cause that's leading to interruptions in your end-user's work day. Don't allow yourself to believe it's a act of God, upper management, or 'just the way things are' because there are too many good shops worldwide that do provide seamless operations every day to allow yourself to accept something less as normal. That kind of thinking comes from relying on the end-users to gauge how well the systems are working instead of on your own knowledge and intuition. Having the end-user serve as the gauge for successful operations means you're counting on them to see the problems before you do. This's as backwards as it gets. If you need non-techies to point out to what's wrong with your systems, then you should do a solid reevaluation of your approach to systems maintenance in general. If you're good at your job, the end-users will never report problems for the simple reason that problems don't reach them. Your attentiveness to potential issues will take care of any problem long before it affects production.

If you want to sustain the professional status of your systems over a long period of time, establish a framework around your daily operations filled with tools that make you the first line of defense against problems. Don't let that responsibility fall to your end-users. If you don't have the tools in-house to support the preemptive fix, then get them. If you need better monitoring capabilities or improved analysis feature sets, then get them. If you need reliable diagnostic tools with automated alerts and reporting, then get them. It these things already exist and it's simply a matter of more vigilance on your part, then step up to the task. Don't allow your end-users to suffer because of a lack of ability to know how well your systems are running at all times.

If necessary, use your communication skills to advocate to management for needed system monitoring improvements. Cost will always be the main concern, but cost is a tricky thing in IT when there are so many intangibles defying spreadsheet entry. Cost in lost time can be an easily calculated when measured against key-strokes, but it's not so easy when trying to calculate the impact of end-user frustration over a marginal computer they're using. Frustration can lead to performance decline which will cost the company money just the same as lost tangibles do. Remember that just because the finance department may have difficulty quantifying the cost of end-user unhappiness doesn't mean you should too. The end-user is depending on you to maintain the tools they use to perform their jobs. The smoother those tools work over time, the happier the end-user will be and the more profitable the company will be as well.

This means any lack of commitment on your part to sustaining the silent system will probably cost your company money. That's true even if the company is already profitable and the systems continually work well enough to make it through each day. Never forget that if those systems cost a single minute of concern to a single employee, then the overall profitability of the company will suffer a tiny measure too. Your work is really that important. Even when your systems are running just fine, you must always focus on keeping operational costs as low as possible. That doesn't mean doing things cheaply; it means doing things efficiently, like a professional. The definition of a well operating system is that it never diminishes an end-user's productivity. It's your job to find the hidden end-user costs in your systems and make them right again.

If upper management refuses to approve the cost of a system improvement you know will help the end-users gain that last bit of operability that gets them to 100%, then make a better case for procuring the improvement and take it back to management to try again. Strive to provide systems that support all the work your end-users do. This includes scaling or even replacing the systems to keep up with new operational requirements. If the finance department needs to see extra IT expenses justified, then exercise your communication skills and build a more compelling argument.

The alternative is allowing the finance department to win too much ground and eventually force you to cut corners in your operations to save money in the short term. This's never a good situation. IT departments will always lose in the long term because the cheaper solutions that end up being used will make the department look unorganized and ineffective as the solutions wheeze along barely getting the job done. What's more, the constant stress of trying to stretch a cheap system's marginal functionality to fit the company's actual needs will keep everyone awake at night. Eventually, the money spent on the cheap solution will be a waste of company revenue and could even negatively impact morale. As a professional, you must find a way to make the case that appropriate funding is always needed for reliable and cost effective corporate IT operations.

If you inherited a marginal system from a previous technician or IT department (e.g. after a corporate merger or acquisition), the rules are the same. You need to convince management that it's losing the company money to keep things as they are. In any case, you must be sure that you provide 100% reliability and functionality to all those end-users counting on you. Don't drop the ball here. Bearing the burden of company infrastructure is what IT techs have been employed to do since the beginning of corporate IT.

Talking with End-users

So how do you know if the end-users are running at full speed with the company's computer systems? The answer is to simply go talk to them. Even if a solution is operating as originally designed 100% of the time, a particular end-user may still be struggling if the system doesn't provide a good fit for a particular task they perform on a regular basis. If all possible contingencies are to be accounted for in the system development stage, this means the best time to talk with end-users about their work is before a new system goes into production. The next good time is when a fundamental shift in company operations about to happen. Again, you should never be in a situation where the end-users are the ones to sound the alarm, so never go 'live' with an incomplete solution, even if the schedule requires it.

It's a different story when a tech inherits a half-baked solution that's already in production. They need to move quickly to collect facts in order to reverse engineer it; otherwise, they can't proceed with a solid knowledge of what needs to happen to improve things. Communication skills are important in this situation too; however, the communication hurdle to be overcome is getting good feedback from all parties who have a stake in the system. That not only includes the system owners, but the system's end-users too. You must be able to talk with the end-users to gather the necessary information, such as how the system is seen from their perspective. Simply asking them a few basic questions won't do it either.

The trick to talking with end-users about their daily requirements is to absorb the information they provide from their point of view and not your own. You shouldn't simply collect some basic answers that you only interpret to your own ends. Instead, you must understand the system from the end-user's point of view to know what their needs are. This change in perspective doesn't come from a few verbal questions or handing out a short questionnaire to be filled in at the end-user's convenience. There are many variables that affect how the information gathering process should proceed. Some of these variables will help you in your efforts and other variables will slow you down. Recognizing between the two which will influence your effort most effectively is critical in the planning of your information gathering approach.

There are some variables that may affect your information gathering efforts. They are:

- **Peer pressure can hamper the information gathering process**

    While it's not the most expedient approach, it's usually best to gather feedback one employee at a time. That's because every workplace has both dominate and submissive people. Both types may have good commentary to offer, but it's only the dominate ones who speak up. If you're asking for input from a group, the dominate types do all the talking and the rest just listen while keeping their own suggestions to themselves. This's particularly harmful to your aims if these dominant coworkers head off on some tangent that only suits their own interests and not yours (or anyone else's).

    Don't let the dominant, tangent-bound, self-servers knock your information gathering off track. If you absolutely must gather your user requirements from a group, then prevent this from happening by finding ways to mitigate the impact of dominant participants. Simply asking, "Does anyone else have any suggestions?" just doesn't do it; especially in any group of individuals when quiet ones won't speak up and yet aggressive brainstorming is required.

    If the group is physically sitting in the same room, using a method of gathering their feedback that puts everyone on individual footing will make your job easier. Plus, you'll end up with a more complete blend of usable information in the overall requirements list.

    A standard trick here is to know who you're inviting to the information gathering session in the first place and try to keep the dominate employees in one group and the quieter ones in another. That way, everyone is more or less in their preferred element. Work with the groups accordingly and don't be afraid to step in and manage the discussion to give everyone a chance to speak. The answers you get will be more diverse and more honest. Honesty and diversity will always give you the best set of end-user requirements you could ask for.

- **Frustration isn't part of the user requirements gathering process**

Some may not understand their job process and want a solution that makes up for that lack of knowledge. Others may just want to complain about any company change under the sun. Don't let either type of input affect your decision making process. It's better to recognize any self-serving user's input from the start and work them off to the side. This includes your boss's suggestions too if they seem a bit self-serving. Whether they're going on about a system they had in another company or ranting about a problem they had with their computer last week, it doesn't matter. Any problems with the legacy systems in your company are something you already know about, so you don't need to hear any re-hashed details. The focus must remain on gathering information for the improved solution and not on collecting gripes about the old one.

Too often gripes can also stem from a person's need for self-expression. If they rarely have a chance to give their proverbial two cents worth of advice on what they see happening around the office, then be careful they're not overdoing their answers to your questions to compensate. The problem here is if you let one person skew the overall system design towards their immediate concerns, others will want you to extend the favor to them as well. In IT, you can never serve the end-users one at a time; you must always serve them as a group. That means staying focused enough to put the company-wide solution before anyone who is expressing an individual appeal. It's the only way to ensure the solution you're designing can be evenly applied across all end-users in your company.

- **Any solution will do as far as they're concerned...even a lousy one**

  Obviously this kind of end-user's input is unsatisfactory. Sometimes the input stems from general malaise about all that's wrong the world. You can't really do much with those answers other than recommend a visit to a mountain top guru to find the meaning of life. With others it could come from a total lack of commitment to the company in general. This is a bit more troublesome. If the person doesn't plan on building a career with the company and is just passing through, then perhaps their advice won't be that useful regardless. Their lack of commitment usually makes them inattentive to their jobs in general and unsuited to providing opinions on something as important as a solution the whole company will use. It's better to let the easily satisfied off the hook and not pursue their feedback. You'll only be mining low grade ore when richer deposits of end-user feedback are available elsewhere.

- **Some may not want to rock the boat by suggesting an idea different from what others want**

  This may be a result of job insecurity and can be a problem in a marginal work environment where employees are worrying about their jobs. Anything they say could make them look bad to management. The best practice here is to reassure end-users who feel this way that their input is totally confidential and needed to develop a system that'll benefit everyone in the long run. Let them know they're important and their experience is useful. Whatever they can contribute will be welcome. If you give them a little confidence, they usually add some helpful input. After all, everyone likes to have a say in things if given a chance. Letting somebody clam up when you know they have good things to say is a waste of human ingenuity. Let everyone know that you don't mind them rocking the boat in your presence. What they have to say is confidential and it's important they speak freely so you can gather a complete requirements list. If they would like a whole other component added to the solution or a very different approach to implementing the solution than what management envisioned, then encourage them to explain it freely. If their input is insightful enough, it could be well worth fighting for when you take the user requirements results back to management for further discussion.

- **End-users' answers may be so non-technical you have to interpret their suggestions into a usable result**

What if someone would really like the new solution if only those 'thingamajigs' could work better? What if they can't explain it any better than that? Well then it's your job to explain it for them. You must cover the distance between their lack of understanding and the input you require from them. Start by ensuring that they have a practical understanding of what's being worked on and how it will affect them. It can't be said enough: the best measure of how well you understand anything complicated is how simply you can explain it to the uneducated. Talk to them on their terms and fill in the gaps with as clear a picture as you can.

Those end-users can be highly educated and intelligent people, just not in your discipline of IT. There's no point in asking the opinion of a person who doesn't understand the context of your question. For feedback to be useful, make sure that everyone has a full understanding of what's being considered. It's up to you to cover this ground. Even if you're talking to end-users who are being dismissive because they're so well educated in their own field they don't need to recognize yours, find a way to break through anyway. They may see your job as menial and want you to do it on your own without bothering them, but get your communication skills going and convince them they have too at stake to not be involved. This will eventually be a solution they work with every day, so of course they should offer some input on how it should work. It's your job to consolidate the sum of end-user opinions into feedback that covers the entire solution, even if no single end-user is able to fully understand the entire solution the same as you do. Whether their answers are technical or not makes no difference. They still offer a value you can use to build and maintain the right tools for them to use.

- **Feedback is important even from people who will only work with a small part of the system**

  While the focus must be on the entire solution, feedback from the limited access end-users will be based only on the little part they see on a daily basis. Treat their input like a snapshot and apply it along with the other snapshots from other similar end-users to the overall context of what you're putting together. Each limited usage snapshot increases your focus and helps to bring a granular clarity that will go far in accommodating their particular needs.

Finally...

There'll likely be a rare occasion in a long IT career where, for whatever reason, you find yourself working on a system you don't think is a very good solution. Regardless, *never* agree with an end-user who claims the solution is bad even if you feel that way yourself. If the system was handed to you to develop or maintain, don't use the end-users in an effort to campaign against it. This's something that must stay internal to the IT department and be discussed confidentially with the appropriate departmental personnel. To talk with end-users about your departmental gripes is immature. If a tech is feeling unappreciated, they may feel inclined to agree with an end-user's lousy assessment of a computer system the company is using. It may make the tech feel better for the moment, but aside from being a purely cowardly act that no self-respecting IT pro would ever do, it is also bad practice. First, if you set yourself up as the tech 'on the inside' who's looking out for the end-users, any problem they see in the future will have the "come help right now or you'll be letting us down" element included when they contact you directly for assistance. Nothing corrupts a busy work day more than making unscheduled house calls for special users. Second, you're setting up your department for derision by inviting a look into what should otherwise be a confidential department issue. There's no need to explain how that should turn out for you.

CHAPTER FIVE

Skill-set Management

While knowledge and experience accumulate only when opportunity allows for growth during the course of a well-traveled career, wisdom is gained steadily throughout. Wisdom is what allows you to make a solid assessment of where you stand in terms of your abilities on any given day. However, expertise comes to anyone who does anything for a long time. It's the result of doing something over and over again until you can predict its outcomes without any more thought. While expertise is important in any career, you must be careful how broadly you define the concept of expertise in the rapidly changing world of Information Technology.

Over-confidence in your expertise can even be a hindrance in a career that requires you completely overturn your skill-sets every three to four years to remain current with the latest trends in technology. Over confidence in past knowledge can fool anyone into believing solutions that were expertly implemented back in the early days, and which may even have won praise for efficiency, will continue to work in new environments where they no longer fit. Those techs will continue to press on nonetheless, convinced they'll make everything work out in the end. After all, they always did in the past. To be clear, relying too heavily on out of date knowledge isn't a slippery slope that will send you plummeting to your demise, but it does cause a slow erosion of your skills. You'll be inclined to stick with old approaches while losing track of the new approaches to technology needed to do a good job today. Techs can lean on their expertise when cutting corners to make *somewhat* familiar tasks easier. Unfortunately, in many cases this can be professional poison.

It's especially true in a career where the only thing that matters is what you learned yesterday, not how much you knew in the past. As far as IT is concerned, it's the wisdom you gain steadily throughout your career travels that'll help you access where you really stand skillwise in the most current settings. Your wisdom will remind you to keep developing new skill-sets on a regular basis. So when, as an IT expert, you've decided you know enough about computers to not have to learn anymore, it's wouldn't be a bad idea to fall back on your wisdom to get you back on track. You've proved you know how to do the job well because you've done the job well in the past. You did it by gaining an appropriate level of expertise in the areas you were responsible for. Your wisdom will remind you that this particular condition never changes. Expertise must always be regained with each new project or system you work on. While the constant struggle to regain a level of expertise can be exhausting over time, you must accept that it's the rejuvenating challenge of constantly learning new things that will support a long IT career.

In fact, learning burn-out is by far the biggest career killer in IT. Every profession requires remaining up to date with the latest trends, but in the world of computers, the obsolescence rate of new technology means any knowledge you gained on a new system will lose its pertinence within as little as three years. At that point, you'll find yourself back at the beginning again with a whole new range of things to learn from scratch. What's more, any certifications you earned will lose their shine too. The worst part is that there are no short-cuts. There's no secret method for sneaking around the brutal learning curve. You can only accept that the demand for learning in IT never ends. This's the path you've chosen as a professional and if you want a long career, you must find productive ways to deal with never ending educational requirements. So it's best to determine right from the beginning of your IT career which learning techniques will keep you technologically current as painlessly as possible.

It can be daunting when you think of all the learning, certifying, and applying of new knowledge you'll have to go through before you can go from what's on the market today to the inevitable release of "Big E-mail Server 2024." As usual, the newer version will be "much better" than the previous version (not to mention every other version before that), so the boss of the future will want it in production by 2025, even if the bugs aren't worked out yet by the manufacturer. However, that's not the truly daunting part. The truly daunting part is that some kid born just ten years ago may end up knowing as much about that new system when it finally hits the market as you do even if you've worked ten years longer in the industry! He'll be fresh out of school and you'll be fresh off the books...same difference. You'll both be starting at the same point as relative rookies. It may be hard to get used to, human nature being what it is, but it goes with the IT profession. When working in IT, everyone will always be a rookie on something. There's no way around it.

Learning all the time is what all techs do. If you take an honest look where your expertise stands at this point in your career and see it just sitting there resting comfortably on its laurels, then you are either waiting to move up to management or you need to start cracking those books again. Otherwise, you're just becoming a liability to your department and to yourself…and not a minor liability either. It's the kind of liability that sneaks up before you realize it. It's the kind that waits until after a mistake is made and things have proceeded past an easy recovery point before it shows its ugly head. It's only then that you finally realize you're no longer ready for the jobs you're given.

Your past successes have convinced you that everything's going to turn out fine when things get tough, but the truth is, you've let your skills fade to the point that they're no longer reliable. That's why good skill-set maintenance makes everything you do easier and the systems you maintain work better. Some skill-sets, like network addressing for instance, can be a part of every tech's knowledge and understanding. Other skill-sets may be specific only to certain jobs in a customized environment. Either way, the more knowledge you can call on when starting a job, the more easily you can move through it.

If you are new to the world of IT, don't let yourself think for a minute that natural ability has anything to do with being successful. As far as the real world is concerned, the moment you decide you're *just plain good* is the moment you become a liability waiting to happen. It's better to remind yourself that continually developing your skill-sets should be a core focus throughout your career. Natural ability, if it even exists, is fleeting. The best approach for sustaining a long career in IT is one that will help you manage constantly updating your knowledge with as little stress as possible. There are a couple of reasons for this.

Reason number one is that you're only as good as your knowledge is up to date. Drawing on past experience doesn't help much in the IT world. Always remember that in IT, you're only as good as what you know about the systems in front of you. You must keep the ability to learn new things at the forefront of your skills. Nothing else defines who you are more clearly than your ability to learn new things quickly. In fact, it's the ultimate compliment in the IT world. You have every right to feel proud when a co-worker or boss notices how good you are at picking up new knowledge. It's the source to all other skill-sets and something you should strive to maintain everyday of your working life.

Reason number two is that when you're accomplished at learning new things, you'll be better able to see that point which comes inevitably to every tech's career: the point when it's time to retire. If you don't continually put in the effort to learn new things as you travel along the career road, the time to stop and retire won't be so clearly defined to you. It's like the aging athlete who won't retire until he has hurt his own credibility and is virtually, albeit diplomatically, kicked out of the sport he used to excel in.

If you are good at keeping up with learning, your decision to retire won't be based on how miserable your career is leaving you feeling every day. Staying on top of your game gives you the best view of when it's time to quit. Anything less will only leave you struggling to understand why the work you once loved is now making you so unhappy. Each new day will seem ever more difficult with no comfort in sight. Always remember there's no better time to bow out of a long career than when you're still in charge of your skills. Being in charge of your expertise will allow you to walk away comfortably from your career instead of running away in misery. Some IT professionals love learning so much that it never gets old for them. With great energy they go on long past an age when many others retire. Then there are other techs who can't wait for that permanent vacation to the fishing hole as soon as they turn forty-five. How you handle your career is up to you, but the best thing you can hope for is that when the end comes, you'll recognize it clearly. The best way to do this is to stay involved with learning new things on a daily basis.

Staying pressed up against the window pane of knowledge gives you the clearest view of when it's time to let go. You'll be able to see your current situation in the context of your life and have a better sense of when the burden of continually learning no longer brings the satisfying returns it used to. However, if you allow yourself the luxury of floating along lazily behind the learning curve, you can become hampered by the uncertainty of what to do with an increasingly stressed and unhappy workload. You won't be able to tell if the lack of satisfaction you feel with your career is caused by an honest desire to go in another life direction or simply the result of you losing confidence in your ability to get the job done well.

It's tough to work when you're worried that those around you might notice you're not as good as you used to be. The more you let your skills slip, the more sensitive to other's perceptions you become. Soon you'll be more involved with hiding your inadequacies than getting the job done. Before long, you'll find yourself being one of the last ones asked to handle the tough jobs when you used to be one of the first. One day you might get tired of being overlooked and decide you'll buckle down and work hard to regain the prominence you once had, but the energy just isn't there anymore and the distance to be covered is just too great, so you continue your downward drift until you end up with no place to work because you don't have anything tangible left to offer. This's a sad way to leave a career. It would have been far better to go out on top. However, the only way to do that is to keep your skills as sharp as they can be until the end. Staying current with your knowledge not only allows you to continue doing your daily tasks well, it also allows you to manage your career, and ultimately your life, better too.

While overconfidence in your expertise is generally useless in a rapidly changing world of IT, wisdom will help you remember how to stay on your career path when your energy is running low and your eyes are getting droopy. While knowledge helps you do things, wisdom can help you avoid pitfalls that can trip you up along the way. Here are some things to consider for keeping your career moving smoothly ahead while learning new technologies on a never ending basis.

Career Pitfalls

The biggest career pitfall is holding onto knowledge beyond its usefulness. In this instance, wisdom always trumps expertise. Every tech who has been around a while is guilty at least once of allocating vital brain space to remembering information long since passed its usefulness. It's wisdom that'll help you keep track of one of IT's golden rules: the moment knowledge becomes nostalgic, dump it from your memory. Being able to tell a newbie how you used HIMEM to move drivers into upper memory on DOS 3.2 PCs back in the late 1980's may make you feel proud, but it only proves you're wasteful. That knowledge should have been dumped long (long, long...) ago to make room for the things you need to know today. If you're stuck in a legacy world supporting out-of-date systems being held in production for regulatory or procurement cost reasons, then that's another matter. However, as soon as those legacy systems are gone from your life, so should all knowledge of how to run them.

Always make room for the new. Holding on to old knowledge is harmful. It's the same as hoarding any other useless items. Out of date knowledge is baggage that'll weigh you down in the long run. This isn't a new concept either; it's where the expression "he's forgotten more than you know" comes from. People who work with rapidly changing knowledge have been deliberately forgetting the old to make room for the new for generations, at least since the start of the Industrial Age. That's because what's important in a world of changing technology is having knowledge that's pertinent to the job at hand and not residual knowledge form old jobs long since completed.

In fact, there are excellent techs with impressive accomplishments from their past that aren't even mentioned in their current resumes. Who cares about some old system that's been obsolete forever? Unless they're citing the experience to demonstrate organizational skills for a planned jump into management, someone striving to achieve excellence in IT would have forgotten about those systems years ago. It's the nature of the tech business; old memories are a burden. What you know is best forgotten as soon as it's no longer needed in your current work environment. If you find one day that something you did in the past is suddenly pertinent again, say a change in employment that involves a move back to an older environment you once knew, then just relearn it. It's easier to relearn things a second time around anyway, but keep in mind this example is an extremely rare event.

The far more likely scenario is that if the world has moved on, your employers will have moved on too and you won't see those systems again. Practical examples of this abound. Take for instance the myth about reusing code to save time and energy when writing something new. Reusing code in an industry that turns itself over every three years only means there's a lot out of date and marginally applied code out there. Snippets of old code here and there might be OK, but more often than not, these shortcuts produce a result inferior to what could have accomplished had the programmer started from scratch. This remains true even if there's a time constraint or a need to blend an old system with a new one. Yes, that was some great programming they wrote way back when, and it still looks good to them today, but the new code they write from scratch today will be even better. As you move through your career, you'll have more experience to use in creating ever better work than before.

The importance of dumping old knowledge is even greater when going over the notes you took while working on different projects. Everybody jots down a few things as they proceed through tasks, it's helpful to keep track of what you're doing, and taking notes as you work is an excellent thing to do in the short term. In fact, that's where all those helpful knowledge bases come from that all techs use. In the long term however, assorted notes from jobs long completed can become a lot of baggage. Old notes are best used for remembering the good old days and should be tucked away in deep storage under after you're retired. Treat them like snapshots from an old vacation. Good for fond memories, but if you go back to visit that place again today, it's probably going to look a lot different and need to be rediscovered all over again. Over time, relying too much on old notes will hollow out your skill-sets and leave you grasping to concepts that no longer apply to the latest systems. In IT, it's better to have a quick mind than a long memory. Notes are good for remembering only as long as they're current to the task at hand. There's no place in a current IT shop to bring up the past unless the past is current enough, say less than three years, to make a tangible difference in work being done today. Teams, especially the young guys, won't benefit from hearing the older guys reminisce about the systems they worked on back in the old days. Keep in mind the world of IT is now well into its second human generation and all those first generation techs are all getting gray hair and long in the tooth. It's important for continuity's sake in our changing industry for older techs to find ways to either keep up with the second generation or keep quiet and fade away. This's the eventual reality for everybody. Those younger techs will be old techs too one day and face the same harsh reality in their long careers.

Keeping Your Skill-sets Current

So the big question remains; how can you keep your skill-sets current without suffering from the learning burn-out that has hampered many IT careers and left less motivated technicians by the side of the road? The answer starts with how committed you are to your own career. That is something you'll need to discover on your own.

What works for each tech is highly personal, however here are three steps you can follow that might help you in this endeavor:

- **Step One is to find out what you like doing in IT**

  Pay attention to what excites you the most among the activities you see going on around you. This isn't a small thing so don't just brush it aside. In the long run it'll serve you better than simply taking jobs you're not so thrilled with just because they pay the bills for the time being. Following a career path that doesn't satisfy you is difficult enough on its own. If you add in the tedium brought on by the never ending demand for constant learning, you'll be burnt out in no time. The more you narrow your focus to those areas that interest you and then follow that focus to the particular area you want to be in, the more job satisfaction you'll maintain in the long run and the less likely that burnout will enter your life. If you haven't done so already, immediately begin looking closely at your daily task list and honestly evaluate if it's as satisfying as you hoped for when you first started out in IT. It doesn't matter what the work is; what's important is whether or not the work is keeping you interested. Whether working on hardware, designing systems, writing programming, securing networks, testing for quality assurance, or any of the dozens of other paths to follow in IT, what you do should always keep you interested.

  Furthermore, whatever IT field you're in, pay attention to those aspects of your particular field that appeal to you most. If you're a programmer, do you like high level programming or low level programming? Would you rather work on video games or write applications for industrial robotics? There are many types of programming, but each may not appeal to you to the same extent. Even within a narrow IT choice, say in computer networking for example, you might find working with routing more intriguing than planning and laying cable...or vice versa. The point is that the more you focus on those aspects of your job that you enjoy, the more opportunities you'll find to do the jobs you like. Eventually you'll build a reputation for being particularly good at those types of jobs.

  The nice thing is that if you like what you do, you'll last longer at doing it. The more interest you have in what you do every day leads to more comfort in your day to day tasks. The more comfort you have when doing those tasks, the more energy you'll maintain over the course of your career. If you move toward your strengths and away from your weaknesses, the continuous new learning won't seem quite as tedious.

- **Step Two is to rely on yourself and not your employer**

Don't rely solely on your employer to gain whatever knowledge you need. Having an employer pick up the cost for your learning is great when it happens, just keep in mind that your education can be costly for companies too. Waiting for your current employer to pick up the cost of an expensive computer course or even a simple certification guide isn't a reliable way to move forward in your career. You should develop a formula for staying up to date with knowledge on your own that works best for you.

It must be a formula which sustains you for a long time without burning you out or breaking your budget. Bootcamps and continuing education courses at a local college can become expensive over time when you add in the constant need to learn new things. Plus the money you paid for courses will lose its value as soon as the knowledge becomes obsolete. When it's time to learn something new, the cost cycle starts all over again. As an alternative, the most affordable path that techs follow is to buy self-study guides that provide certification level learning for whatever it is they're working on. This means you become your own teacher and, as such, requires the self-discipline to set up your own study schedule and stick to it religiously.

You also need to determine how much learning you actually need. Are you working toward testing for an engineering level certification or are you just planning to learn enough to successfully complete a particular job you've been assigned? Measure this carefully because learning, especially when certification exams are involved, can be a lot of work and, as a result, a prime source of stress. Not all learning needs to lead to a certification. Best practice is to keep engineering certifications a personal goal since they'll travel with you as you move from one employer to the next. On the other hand, company paid education is usually niche oriented to a specific system and not very portable to other employers. However, if it helps you work in that niche better, then definitely go for it. Even if it doesn't lead to certification, it's one more important thing you'll know about and will make you better on the job. Your good work ethic always follows you more closely over the course of your career than any certification can. Certifications fade away with age, but excellent work habits will never be obsolete.

Occasionally a contract may require your company to have a certain percentage of techs who are manufacturer certified on a particular solution. In fact, this is commonly the case in government contracting. However, unless a business opportunity comes along that requires a certain number of certified techs on staff, there's a chance your employer won't be interested in even covering the cost of your certification tests, let alone the costs of the courses. If your employer will pay for your training and certification, that's wonderful. If they just pay for your certification testing, that's not so bad either. Just don't wait for this type of reimbursement policy to go into effect if one doesn't exist.

The need to keep abreast of new technology in IT won't wait for policy decisions by company management. As far as the costs for certification tests go, the usual formula IT departments for expensing study requirements is to pay for the first certification test attempt, pass or fail, and then any successful attempt after that. If you fail the test more than once, you pick up the bill on those extra results on your own. However, you may often end up paying the entire bill for learning materials on your own unless you can convince your boss the learning is operationally critical.

While the ability to expense the cost of learning to your current employer is certainly an excellent thing to be able to do, what you want to avoid is letting the people you work for have too much say in the direction of your career. If you only train when required on systems your employer wants you to learn about, there's a chance you'll only be useful to that particular employer over time. Try to avoid this since it could cost you your job mobility. Never think it's excessive to go out on your own and train on something your company doesn't necessarily need. Your company may not need it, but that knowledge could benefit you long enough to help you walk a career path more to your own liking. A self-serving approach to knowledge building helps you to find that particular niche in computing that interests you the most. If this approach leads you to moving on to a different employer who's offering an opportunity for you to work more closely with the tasks you like, then so be it. You'll be a better match for the new position so both you and your new employer end up winning. This's far better than continuing to underachieve for an old employer by sticking with work you've lost interest in doing..

In all cases, learn how to work around the cost of learning. The typical certification process can be too high-priced for most people. It can easily make the average tech's continuing education too costly for anything that isn't absolutely mandatory. This's especially true when your employer isn't picking up the bill. If it's too expensive for a company's operating budget, then it's likely too expensive for your personal budget. When you take into account that you need to continue learning throughout your career, budgeting your learning dollars is critical. Always try to get the most out of your hard-earned money. If you can manage to occasionally afford to attend courses, then that should be your first choice. Courses tend to be well structured and compact and they avoid some of the distraction and inefficiency that can be part of the self-study approach.

However, the reality is that the most common learning method you'll be able to afford throughout your career will probably be self-study training. It's simply the most cost efficient way to go. Because of this, as soon as you can you must figure out how to learn on your own in a way that work bests for you. Keep in mind that learning is different for everybody so choose what actually works best for just you. Your coworkers may have lots of great suggestions for learning material, but ultimately it must be your call alone. When you find the best path for renewing your knowledge, you'll be able to learn for many years with a minimum amount of stress. Find the method that works best for you personally while keeping in mind you'll be paying for most of it.

- **Step Three is to find a learning source that fits you the best**

No two techs are alike so, even if someone else has great advice about study materials they like to use, never lose touch with your own sensibilities on the matter. There are many sources available for your own particular style of learning. These include certification books, vendor material, exam preps, admin guides, online brain dumps, and even hands-on experimentation when you can get it. It's critical to a long and happy career that you get an early understanding of what works best and then create a library of resources as your career and knowledge requirements grow. Having a place to go for the results when you need to learn something new is a great way to achieve, and sustain, excellence in what you do. Otherwise, you might end up perpetually experimenting with different resources and getting mixed results that cause stress and ongoing performance issues.

Usually the most accessible learning tool available will be a certification guide book or series of certification guide books. They tend to have the most information concentrated into the smallest space and so are perhaps the most convenient, as well as the most affordable, method of learning. Of course, it's not simply a matter of reading through the hundreds of pages of technical data the guides provide. You must also memorize virtually every important fact in the guides too if you're planning to take a certification exam afterwards. So it's important that you find a publisher that presents information in a format that agrees with your personal tastes. Some publishers present topics from a real world perspective. Others focus only on getting you successfully past your certification exam. Some are dry, but direct and a few can be light and easier to read, but can get a bit windy. Many include some type of self-assessment tool such as practice exams you can use to check your progress. All are helpful at knowledge building so it's up to you to find the publisher whose materials work best for you.

Of course if you buy books from a publisher for the first time as an experiment to see if their format works for you, and it doesn't, it may seem like a waste of money. Nevertheless, it's the only realistic way to go when it comes to finding your preferred publisher. There are the occasional sample chapters you can get from online book promotions, but these are usually incomplete and fail to give you a real sense of how well that particular publisher's approach will actually work for you. Until you settle on a publisher you feel meets your needs for the foreseeable future, there'll always be some measure of experimentation with the guide books and other materials you buy. There's just no way to get around it.

The most important characteristic to look for in a certification guide is readability. This isn't just a measure of how accessible the language of the author is, it's also a measure of how well the information presented works with your learning style and environment. Not all books are the same for everybody. If you have a lot of hands-on tools around you, for instance a bench environment at your workplace, then to-the-point manuals that have lots of exercises might be good. However, if you don't have the ability to try stuff out while you learn about it, then a more holistic approach that describes concepts as well as the steps to follow might be better. Choose whatever approach works best for you.

When you finally settle on your material and start studying a new technology, it's important to remember one of the cardinal rules in the world of IT: never try to learn about something new by working on it in a production environment. Better still, repeat the following sentence out loud: "I (insert your name here), promise to never try to learn a new skill-set by experimenting with it in production." The reason is simple. If you break something you don't fully understand, then you aren't going to know how to fix it either. Either vendor tech support or a more knowledgeable coworker will have to come to your rescue. The temptation to tamper with a production system can be strong, especially if the program or technology you're studying is running right there in front of you. Avoid the temptation at all costs. You must be professional enough to wait until you fully understand what you are doing before attempting to service it. If you don't have a training center in your office, then find a way to build a system or two on your own that will allow you to practice safely.

Ways to Learn Well

Boot Camps

A common source for learning are the commercially run crash learning courses often referred to as "boot-camps." These usually last three to five days of classroom time and are meant to provide an immediate and highly structured path to knowledge over a short, albeit expensive, period of time. If you in a rush, they're considered an improvement over the slower paced instruction provided by Adult Learning Continuation centers or local colleges. However, that extra level of intensity comes with two problems. First, boot-camps tend to be expensive to the point of being cost-prohibitive for the average work-a-day tech. Second, they only work when followed by immersion. That means either use that new knowledge right away or you risk losing it. The faster you can put all that new information crammed into your head into use, the better return on your investment you'll have. The crash course comes with the caveat that as quickly as you learn, you can forget too. Be sure to enforce the new knowledge you've just gained through the solid practical application so it can establish a long term presence in your mind.

If you've spent a lot of money on a boot-camp but, for some unforeseen reason, aren't able to put new learning to immediate use after you graduate, then find a way to retain it, short of playing around in production of course. Either set up a practice system or do a lot of reading for reinforcement. Otherwise, your expensive investment will fade away before you have a chance to enjoy its benefits. Boot-camps are great if you need to become proficient on a new skill-set quickly. They're better still if your employer is picking up the cost. Just be sure you schedule it properly so that you can hit the ground running as soon as the course is over.

Self-paced Learning

It's possible to say the same about self-paced book-learning too, but with self-paced learning, it's more about the interest level. If the chance to use your new knowledge is delayed for some reason, you can simply go back to the learning guide to refresh. This's where going for an industry certification instead of just gaining knowledge needed for current task can be a better in the long run. An industry certification gives you a long term study goal separate from the goal of just completing a job related task.

However, even with long term goals you must still schedule work properly. Going back to the same learning materials to stay refreshed until the knowledge is eventually needed can become tiresome. It's no different than reading any other book over and over. Pretty soon you become so tired of the book you won't want to read it again. Especially if it contains a thousand pages or more like many guides do. You should also make an effort to get access to a practice environment to keep your new skills sharp until they're finally needed.

Pay Tech Sites

Pay tech sites can be another good source of support for learning, but be careful. You don't necessarily get what you pay for with "expert" web sites. Pay sites are sometimes no better than the free tech blogs. However, good pay sites that offer membership and the ability to both contribute and learn are a real asset because they include expert columns and how-to articles that rise above the usual tech chatter. They also provide some direct application assistance that goes beyond the vendor's level of support, such as information gleaned from how the system actually works in the real world and not just from how it behaved in a manufacturer's development lab. Of course, pay sites aren't free, so shop around a bit. You may occasionally waste money on a crappy pay site, but it's worth the investment. In the end you'll have found both a source of information you like and the positive feeling of belonging to an IT group that cares about being useful.

There are also free methods that can also be good for learning…in some cases. However, while "you get what you pay for" holds true for many of these resources, many free sources of information aren't bad at all. They can come in the form of online resources, such as knowledge bases, blogs, and brain dumps. They may also be something as immediate as another tech in your department who already works with the system and wouldn't mind passing along some helpful information.

All blogs however must be taken with a grain of salt. Generally speaking, the more "chatty" they are, the less helpful they'll be. Stay away from sites that allow opinionated techs to vent. They'll not only be a waste of your precious learning time, but are usually frequented

119

by a lower grade of tech who gives marginal advice even on a good day. Participating in these sites with the hope of raising the level of discourse over the rancor usually has no impact and can expose you as a target of some of the rants. No good can come from inviting others to take shots at your confidence. Computers don't have strong emotions and it helps if technical discussions about computers don't either. If you detect emoting from contributors, it's probably best to move on to another site where information sharing doesn't include venting. Just remember there are other sources of information out there besides free information exchange sites.

Learn by Doing

A pretty basic and fun tactic that IT pros use to learn something brand new is to just start working with it in a safe place. This includes everything from software revisions to complete operating systems. Of course it can't be done anywhere near a production environment, so requires having accessible hardware around that's still new enough to support the solution's specs. Unfortunately, a training system isn't an affordable option for most techs and is often out of the reach in terms of cost. This's where employers come in (and any good employer should already know this).

The best way to maintain a skilled workforce is to set aside a learning area somewhere inside the department or bench area. It doesn't have to be fancy, just a room with some tables and chairs, some spare computers, servers, some spare peripherals for networking, and sufficient power to run them. This isn't a tech lounge for wasting time, but a place where IT staff can go to whenever they have a little downtime and want to get hands-on experience in a safe environment.

This learning environment can be shared by all techs individually and also support small teams looking at a new solution. It's all about practice. In any performance oriented environment, say sports or the military, practice and repeated drill is essential to maintaining a high level of output. This concept applies in the Information Technology world too. However, in IT, it also includes a certain measure of experimentation as new ideas are tried and new solutions are learned. The nice thing about a learning center is that it gives techs the freedom to build and destroy without hesitation. There are few better ways to learn about a system than by repeatedly building and destroying it until its essence is second nature to you.

For example, an excellent way to learn a new flavor of operating system is to simply install it over and over again while adding on one more new feature each time. Take Linux for instance. A tech could buy an installation CD or download an open source installation off the internet. They could start the installation while referencing the man pages that came with the software. When done inspecting their work, they wipe it out and install it again, this time adding some extra drivers. Then wipe and reinstall again, but now adding some networking connectivity like Samba. The whole time they're referring to the man pages for each new installation. Then they build the OS again, adding an office suite, and yet again while adding task specific applications like Snort or Squid… and on and on for a dozen more rebuilds. The tech continues reinstalling it until the new OS becomes very familiar. You can be workably proficient within a matter of days using this method and no expensive boot camps were required, providing the required resources are available to you.

Listen and Learn

Learning can also be gained from working with others around you. The antidote "helping others helps you too" is not just a campy expression; it holds real value in a world where constant learning is a never ending requirement. Sharing the load of learning with someone else can diminish the burn-out factor quite a bit. Same as in any other stressful environment, support groups are helpful in IT too. Plus, the act of explaining something to someone else is a great way to reinforce it in your own brain. It seems to imprint the information more clearly into your memory. Generally, self-learning in small groups usually works better than learning alone, so take the opportunity to learn in groups every time you can.

Knowledge Imbalance

Sometimes when your new knowledge finally gets put into practice, the implementation of the system you're working on is limited to such an extent that only a few of the new skill-sets you just learned are actually needed. This leads to a knowledge imbalance which can trip you up in the future if you move from one environment containing the system to another one with the same system applied a bit differently. This means you have to review the information again to expand your ability to work with it. Nonetheless, you still can't use this as an excuse to play around in production even if you know the system pretty well at this point. Rather, it's best to find an outlet that allows you to relearn those different aspects of the system you weren't working on earlier. Don't let your past knowledge weigh you down. Approach the new implementation from as fresh a start as you can muster.

Retaining Knowledge

With the requirement for constant learning being the hardest part of a career in IT, don't let yourself waste any hard earned knowledge along the way. Once you've gained new knowledge, find a way to hang on to it even when only a little is needed each day. Keeping your knowledge fresh is an ongoing challenge in the world of IT so knowing when to let go of hard earned knowledge to make room for new is a tough call for any tech. You might let go of information no longer required at one job, only to have it turn up at another job down the road. Your resume may say you're still an expert, but your conscience tells you otherwise. You may push on hoping to regain the knowledge before anyone realizes you're no longer the expert they hoped for, but this can be a gamble and a source of stress.

However, stuffing more information into your already crowded brain isn't always easy. It means regularly letting go of old knowledge to make way for the new. As said before, hanging on to old knowledge too long is counter-productive. It takes up valuable brain space and can burden you with preconceptions when taking on a new job. Old knowledge can cause major issues when trying to figure out a new problem. You could easily presume that what went before still applies today, even if it doesn't. Old preconceptions are a main reason for letting go of knowledge on a regular basis. However, releasing old knowledge doesn't mean burning the bridges that can get you back to it in a pinch. If you need to re-immerse in the technology down the road, it's easier if you have all the necessary materials readily available for recovery and reuse.

It's best to find a way to leave yourself a trail of bread crumbs when you're about to leave some knowledge behind. That way, you can recover the knowledge from your personal archives if it's ever needed again. IT departments can do this too, but it's up to you to look after your own needs in the long run. Bread crumbs can include a certification guide book collection, printouts of articles from old knowledge bases or other sources, personal notes (as long as they're not too systems specific), manuals and how-to guides, and old software installations you still have (as long as it's not a license violation to use them as a personal refresher course). Keep all your resources in a location you can get to fairly quickly and then measure its usefulness over time. Manage it well and eventually move items to the nostalgia bin to make room for the next batch of old stuff that comes along.

Mind Your Health

One last thing regarding the stress of continually gaining new knowledge only to watch it become obsolete is to mind your health. Do not skimp here. Your health and your ability to continually learn go hand in hand. Research has shown that physical exercise reduces the effect of stress; a motivation killer and the leading cause of burn-out. It also helps keep the mind more responsive when digesting new information. The brain benefits from constant use and will stay healthier longer if you keep it exercised. If you can avoid stress, a career in IT has a reciprocal effect on the brain by keeping it constantly honed like an athlete's muscles. The stronger it is, the longer it can keep learning.

Diet can help here too. Even though it may sound a bit preachy, eat healthy and get plenty of sleep and exercise. The last thing you need is a chronic lack of fitness to catch up with you and make your brain less responsive. Train both mentally and physically as with any competition, except in this case, the competition is with computer systems that never sleep and continually morph into something completely new just when you're done understanding them. You have to train for the chase. That includes following the boring advice your mother told you about eating vegetables and drinking your juice. If you want a long and successful career, don't let your health become an issue if you can avoid it. Always seek ways to keep your edge both physically and mentally and the learning you undertake will stay sharp and quick.

The same goes for any tests you may have to take to prove you know the new information, such as those difficult certification exams. Some techs may even be required to pursue certification whether they want to or not by employers hoping to use to the certification for a new contract being bid on. That situation is always about as stressful as learning can get. If the tech fails the exam more than once, they could jeopardize their company's success and damage their career in the process. It's important for every tech to learn how to take exams.

It's also one of the most personalized skills a tech will ever develop. Everybody is different. For example, it may involve aiming for mid-morning test time, drinking a couple (or four) extra cups of coffee, doing one last blast of review in the parking lot before walking in, and then taking several minutes to brain dump every bit of that review blast onto those blank sheets of scratch paper they give you before it fades from your mind. No, you can't enter a test center with written notes, but you can cram a ton of last second facts in your head and throw them on paper the moment you take your seat. Then you take a few long breaths and press that inevitable start button.

When it's all over and you pass the exam, you earn some happy down time for a few months, or even a few years, but then it starts over again...and when it does, it'll be every bit as tough as it was before. The skill-set chase never gets easier as you get older. It'll always take too much time and energy and will leave you wondering if it was truly worth it. When you feel your endurance giving out, have the good sense and professionalism to step aside and let younger minds take over. Good for you if this means moving into IT management, but if you leave IT altogether to try something new, that's good too. The bottom line is that no one should be unhappy with what they do for a living. Just remember that if you like working on computers, then successfully managing the learning demands that a career in computers requires will keep you happily working on those computers for a long time.

CHAPTER SIX

Team Membership

Everyone working in IT needs to have good team membership skills. This goes for big departments, small shops, and even techs who are working in a one person network. You might occasionally be the lone-wolf hero for the day, but to remain effective over a long career you must work well with everyone around you. That means understanding that a team can also be comprised of outside sources such as knowledgeable contractors and quality vendor support staff. Team membership is essential for maintaining a high level of productivity over an extended period of time. Whether you're sharing the bulk of the work load or just sharing concepts at the beginning of the planning stage, everything goes easier when you work with others rather than simply work alone. In fact, the quality of your team skills can be measured by how many others around you have learnt to rely on you on a regular basis. It'll be the people you work with who take the most accurate measure of how good your team skills are. You should make a point in your job to contribute to the group, as well as encourage contribution from them, at every opportunity. Your whole career will be easier if you do.

Listening

Working in a team begins with being a helpful listener. A helpful listener is someone who's willing to act as a soundboard to someone else's ideas or issues for better or worse. Helpful listening is especially important when many people are working on time-sensitive jobs that hit the occasional rough spot every now and then. Listening to your fellow techs describe the situation they're working with helps them to think more clearly through things like difficult trouble calls or an upcoming project milestone they're assigned to complete. Helpful listening means volunteering to be a soundboard to bounce ideas off. It's through helpful listening that you can make a contribution to the other person's thought process.

The act of discussing a problem out loud has a great effect on the mental process. Lots of great ideas throughout history arrived through open-ended argument, scientific debate, discussions over mugs of coffee, or even from reviewing notes with colleagues. Yes, Newton is an exception to the rule, but it's rarely the solitary thinker huddling in a closed room that creates a great work of progress. The average person benefits from getting their mouth working along with their brain to synchronize an assortment of complicated ideas into a practically applicable format.

It's called cognitive semantics and is part of the everyday thought processes we all use to turn complicated concepts into manageable ideas. For the brain to turn ideas into speech, it must first organize and review those ideas and then put them into a format that's more workable at a real world level. Speaking ideas out loud is a helpful tool with problem solving. The end result is uncovering a clearer and more appropriate solution to the problem at hand than one would find from a solution shaped purely by their imagination.

Whether or not more brain power is gained by the speech process, the act of talking by itself has a positive effect on problem solving. Letting your coworkers talk to you about a particular issue, even when you're not all that interested, helps them to work through details that were not so easily lining up in their brain. Offering an idea to them, even one not entirely useful, will give them something to consider that either reaffirms their original train of thought or propels them to a more productive one. This works both ways, too. If you're the one doing the talking and another tech is listening, then their suggestion can help you get a better grip on a problem, even if their suggestion isn't all that good. The act of figuring out why an idea isn't good can often sheds as much light on the correct path to follow as helpful ideas do. Any thinking about a problem, whether working through a bad idea or a good one, will help you to get closer to the solution you need.

The act of thinking verbally is how tough solutions eventually reveal themselves. However, when the right next step is discovered, it's important to remember what was said along the way. This doesn't mean constantly jotting down notes as you talk, but it does mean holding on to those new ideas you discussed in case they can be applied later. If you're able to review the conversation at a later time, there's a chance that even more new ideas could pop into your head. Talking openly with others about the jobs you're working on will help you find the best course of action to follow nearly every time.

Luck Doesn't Count

Another way to be a good team member is to never rely on luck to get something done. A lucky fix which can't be repeated by you can't be repeated by anyone else either. Tough problems are always better solved when you take time to work through each step involved. Getting lucky doesn't count. If you did a tough fix once, you should be able to do it again following the exact same steps as before. If it was a lucky fix, then that won't be the case. Therefore, if you ever find yourself getting a job done better than expected then you should stick around long enough to figure out *why* it turned out so well. Do this no matter how rushed you are.

How completely you do your work is a good measure of how complete your contribution to the team is. The more complete your jobs are, the better the work environment for everyone around you. Haphazard work is a problem for everybody, regardless of how well the job may have managed to turn out in the end. You should be diligent in everything you do and document your course of action so that teammates won't have to spend time at a later date figuring out what you did to get something working correctly. Nothing leaves more unknowns than good luck. Especially when working in a hurry on complicated computer systems. If you rely on luck at all, you're just setting both yourself and your team up for hard work in the future.

Luck becomes a major problem when all those overly fortunate outcomes start adding up into a mess of poorly understood solutions scattered around the network. The company's entire IT infrastructure could become a patchwork of barely maintainable systems and applications. Before long all you can hope for is continued good luck to keep everything running like it was before. It's far better to leave nothing unclear when you finish a job so you can do it exactly the same way again. Make sure you know precisely why something turned out the way that it did. It's not enough to be successful for just one day in IT, to achieve excellence over your career you must know the details of your successes so you can continually repeat them in the future.

Volunteering

An excellent way to be a good team member is to volunteer for new tasks on a regular basis. As long as you're not reaching too far beyond your skill-sets, there's no reason to not step forward on occasion and take on some extra workload. You'll help your teammates and also gain both experience and credibility, two things that go a long way toward having a reputation for good workmanship. The thing to remember is that the opportunity to gain useful experience in IT may not be offered to you as often as you might hope. What's more, those opportunities may be only around long enough for someone else to step forward and claim the chance to work on something new for themselves. If you only wait for new job experience opportunities to be assigned to you, then you'll grow in your career at a much slower pace than volunteering techs will grow in theirs.

When there's an opportunity for you to make a contribution beyond your normal work load, the immediate credit and experience you gain from your willingness to take on the job is something you can carry with you long after the job is done. If you get the job done well and on time, it's a win-win situation for you and your team. It also helps you build a reputation as the tech everyone goes to when all else is failing. Volunteerism is the shortest path to becoming known for excellent workmanship. If you want an excellent tech status, then step forward at every opportunity and establish a reputation of being someone who can handle many jobs well. Just be sure you don't overreach and your career will grow at a rate that keeps you both busy and happy.

The other side of this is the team member who never volunteers. Either they're uncertain about moving forward, or worse, think those who volunteer are easy to take advantage of; a place to hang their workload. The unfortunate truth is that this type of tech is going to eventually watch their co-workers leave them behind. It's a sure thing. The guy who sits back and lets others do the work is always the tech who fades in prominence as his underutilized skill-sets dry up. When the day comes that the under-achieving tech finally decides to volunteer for something, they risk discovering that no one may want them around anymore. After all, theirs reputation says they'll only do as little as possible. No one needs that kind of attitude on a team where everyone else is striving to do good quality work.

Picking up Loose Ends

Volunteering isn't just for when management puts out a request for someone to handle a job. The act of picking up the occasional loose ends missed your coworkers is the most important type of volunteering there is…it's the type that no one gives you credit for. In fact, no one may even know you did the work at all. The only thing anyone might notice is that the systems you work around always seem to run a little better than the rest. It's best to handle loose ends as soon as they're discovered by doing the work without being asked to, even if those loose ends weren't left by you.

Loose ends are defined as a relatively small bit of work that can be overlooked in the scale of a single job, but when compounded by many other loose ends, eventually slows down operations. They can range from unmarked patch cables, to poorly notated scripts, to skipped software updates, and so on. There are many ways to leave loose ends lying around a busy shop and none appear significant by themselves. It's when those loose ends start adding up that the problems begin. IT operations full of loose ends will always be out performed by operations that have none. Leaving no loose ends is a true measure of how good you are at your job. Picking up and completing these little tasks, even when they're not your own, adds greatly to your team's success as well as your peace of mind.

Even good techs might leave a loose end once in awhile. They could be struggling through some tough time management issues and doing work too quickly to be thorough. Perhaps they were in over their head for the job or were too distracted by outside influences to keep their focus. They could even be a reliable tech who just had a bad day. Regardless of whether or not you know who left the loose end, it's usually best to bear down and fix it yourself if it's right there in from of you anyway. If they leave a lot of loose ends then you can mention your contribution to their cause to them in the hopes they clean up their act. Regardless, at least you'll know the work is done and that loose end won't trip up you or anyone else later on when things get busy.

A willingness to make unsung contributions is a critical part of any good team environment. Teams with a good "loose end hygiene" always outperform those who leave systems unkempt. If a tech feels that helping with loose ends is beneath their bearing, they should remember that no one wants to be a big star on a team that botches a job. There's no personal gain in that. If everyone on the team does the small things to keep operations running smoothly and the team moving forward, then everyone one on the team looks like a winner.

Mentoring

Another important component of team membership is mentoring. If you see a tech who's working hard but struggling a bit, help them out. Don't do their job for them, but instead find a way to help them increase their abilities so they can complete the tough tasks they've been assigned. If they're trying but struggling, just think back to those times when you were in the same situation. Those days did exist once. They've existed for every IT professional in the business at some point. With the turnover of IT knowledge happening as such a fast pace, those days will probably exist again for every tech in the future too. Besides that, which individual gets assigned to take the lead for a particular job will probably change as often as technology does. The more each knowledgeable person's expertise is distributed throughout a team via mentoring, the more helpful it will be to everyone on the team. As knowledge grows, so does a young tech's confidence and his/her ability to make meaningful contributions to everyone else's work.

A small boost in confidence can be a useful source of energy for people who feel a little embarrassed about their inability to do jobs they see others completing with ease. If you help a struggling coworker understand how to do a job more ably, then that's one less concern you need to worry about in the future. Their contribution will give you one more asset you can count on when your own workload starts to grow. Team building is always done from the inside, so mentor as often as you can.

On rare occasions, despite the clear benefits of mentoring, some IT pros may experience a manager who thinks that all they need to do is to keep firing and hiring employees until the revolving door creates some kind of dream team of knowledgeable workers. Mentoring doesn't work in that shop because the manager expects everyone to know everything from the start. Don't buy into this managerial style, even if it puts your job on the line. In reality, the manager isn't behaving in a qualified manner for their role and, to be blunt about it, anyone who was ever fired by a fool has never regretted it. This last point is especially true in IT. Still, keep in mind that one bad manager doesn't necessarily ruin an otherwise great place to work so you could also decide to stay put and stand your ground if it suits you. If that's the case, then find ways to confront the manager who doesn't have patience for mentoring or good teamwork skills. This doesn't mean you should argue with them, but you should try to find out what can be done to help them become better at what they do. In a sense, mentor your manager too.

Bad managers usually tend to be touchy and often appear overly frustrated. After all, it's not easy being bad at what you do for a living. That's not a facetious notion either. If you put yourself in their shoes, it's easy to understand how exhausting it is for someone whose work seems to be a never-ending struggle. The standard excuse bad managers often turn to is that everything wrong with their department is the fault of a handful of employees. It's a quick way to dump the stress of their managerial performance issues onto their charges. Trying to mentor others under a bad manager could be seen by the manager as you simply trying to usurp their authority, so techs in this situation prefer to keep quiet and work as individuals rather than as a team. You must remember though that while bad managers come and go, your career will stay with you as long as you want it to. If you're in a workplace suffering from bad management and you don't want to quit, then it'll help your career in the long run if you find ways to communicate with the manager and help them improve their own performance.

If you can find quiet ways to help out, you might find your manager isn't quite as onerous at you thought. Especially once they get a better handle on their situation. You might be amazed how many ineffectual managers know their faults but are afraid to admit it. Bad managers actually do appreciate a little friendly advice on matters every now and then…as long as it's not too outspoken. Occasional suggestions about how a project might proceed or how work performance could be boosted will be heard by such a manager, even if they don't acknowledge it at the time. Before long, you'll see some of those suggestions surreptitiously enter your work environment. Just be sure you're helping the manager without alienating your teammates by appearing to be "in" with the boss. If you stay in touch with your teammates, you can negotiate a political balance between mentoring the boss while still watching out for the workplace. Sound difficult? It is…but if you want to stay at a good company while working for a bad manager, you might have to give it a try.

Setting Goals

Goal setting is a critical part of pretty much anything you do in IT. When working with a team it's important to maintain two separate sets of goals along the way. The first set is the joint team goals, those that clearly define the job's milestones and the job endpoint for everybody. The second set is determined by you alone and reflects your desire to contribute to the team. There's a reason why you need two sets of goals.

Only working toward the team goal makes you an "inbox-outbox" type who performs only specifically assigned tasks with no meaningful input on your own part. You don't actually contribute; you just do what you're told and then go home at the end of the day. This approach may help you remain competent enough to keep your job, but after a while, your fellow techs will learn not to expect much extra effort from you. You only give what the task list outlines for the day and nothing more. Fortunately, most techs *do* care how their contributions are perceived by the rest of their team and genuinely want to do more to get the job done right. If you're in the latter group, then you should maintain a second set of goals that are separate from the team's goals. These are goals you carry with you everywhere you go and will ultimately define your excellence.

You can establish whatever goals you feel confident will help you be successful. A few recommendations are:

- **Laziness is never a positive**

   You can never be both lazy and good at the same time. Be careful because laziness can be a tricky thing too. One example is a senior tech's relaxed approach to new technology because "he's been there before and it's the same old stuff in a new wrapper." Or it could take the form of an old pro's desire to let some younger tech do the grunt work that he'll just polish up at the end. This may seem a reasonable position for an old pro to take, but even old pros can be bad techs if they let themselves get lazy. In fact, laziness that masquerades as dismissive confidence is the most common kind around.

   The old expression "close enough for government work" is usually said by a tech who wants you to think they've got all the parameters and possible outcomes understood as if it were child's play. The truth is that working with computers in a government setting is actually pretty precise and demanding. If you work on government IT jobs, you probably already know that the "close enough" attitude usually doesn't last past the first day on the job. Some government contracts even require that service vendors begin *paying the contracting agency* an hourly rate if they go beyond schedule and the work runs longer than allowed. Those extra hours are billed directly to the vendor as an accounts payable to the contracting agency. In short, if there was ever a career where "close enough" is always a bad idea, it's IT.

Computers are precision machines. Only the lazy think that "close enough" has any measurable value in daily operations. If commercial sector systems were truly running "close enough for government work," many would be operating at a higher level than they do today.

Laziness is underachieving in all the little ways that are part of being a good computer professional. One culprit here is certification. It's been said before in this book; anybody can burn out on industry and vendor certifications. They never get easier. What's worse, after a few years, you'll find out just how fleeting each certification is. It's not unusual for older techs to have an assortment of old certifications for legacy systems nobody uses anymore, made by companies that no longer exist. That's no excuse. It's just one of the paths you have to travel to remain competent in a rapidly changing world.

Get back to those books no matter how painful it may be. Otherwise your great experience and work ethic could blind your respectful co-workers to how little you really know anymore. This can be a liability when you start making mistakes that no one expects. When this happens, there's a lot of explaining to do by the old tech who allowed his/her skills slide a bit too much. They'll need to convince teammates that their skills are still trustworthy while trying to understand how they allowed themselves to sink into that situation in the first place.

Another symptom of laziness is losing a sense of achievement. Problems happen all the time to regular people, but it's different when they happen in the IT world. In

IT, issues are only caused by a lack of overall effort. Professionals know that many computer problems are the result of conditions produced by an underachiever who made an otherwise solid system vulnerable by allowing luck or fate to enter the equation for uptime through a lack of effort. This's forgivable if the offending party was struggling with new technology they aren't quite ready for. At least that makes sense. However, what won't make sense is if the offending party is an experienced IT pro who let their once reliable work ethic fade away through carelessness.

In fact, there was once a tech who actually created a philosophy based on underachievement. It centered on developing maximum likability with his end-users. In truth, he was extraordinarily busy having been assigned maintenance on two major contracts with several sites scattered over a large area. However, instead of working on a better argument to upper management about how his situation was untenable, which it was, he developed the concept that if the people at the sites really liked him, they wouldn't mind living with constant bugs he left in the systems so much.

So his first task when he arrived at each site was to go around glad-handing everybody in the office, asking how their kids were doing, talking about the weather, talking about sports, and so on. Eventually, he would do a quick fix for whatever caused the service call and then head down the road to the next site. This approach worked fairly well for a while...as long as the bugs weren't too annoying. Of course, what he called his "sell the PC approach" (yes, he actually named it) eventually caught up with him. He was certainly well liked at his sites, but those end-users had busy workloads too. After a while, the bugs were causing enough interruptions that complaints eventually percolated up to Executive management and audits were performed. The end result was bad all around and affected the contracting company. Major customer relations work was necessary for both contacts to be saved.

The moral of the story is that the only acceptable outcome to any service call is a bona fide repair. If you can't do an honest job, then it's your duty to yourself, your team, and your career to tell somebody and ask for help. Otherwise you're just being lazy. Never just schmooze with the customer until you can sneak away with a partial fix. This philosophy always fails in the long run. Don't let it become the standard practice. Marginal results will never equal good results. "Close enough for now" says more about the tech than the task. Anyone can do a crappy job. No matter how much time, or ego, or attitude, or whatever you think you'll save by allowing yourself to be lazy, it'll always just a fraction of how much credibility you'll lose if your laziness ever gets your team in trouble.

- **Never walk into a bad situation that you saw coming**

   Not walking into a bad situation you say coming means you should always stay involved enough in the team's activities from the start to have a reasonable amount of input regarding how the job should proceed. Try not to sit back and let others do your thinking for you. If you see something's wrong at the beginning and you don't speak up, you might only be adding baggage by speaking up later on. This's especially true if you knew better than the others did but let problems happen anyway.

   Of course, new ideas have a way of popping out of nowhere while a job is in progress. Whether it's an act of fate or an outside force exerting a necessary but unexpected change requirement, paying attention to things as you go can jump-start the creative juices needed to develop new approaches and angles of attack as the situation develops...as long as it doesn't interrupt the well thought-out project plan of course. However, being reactive won't do any good if you don't speak up when the opportunity arises. Letting someone less reactive than you steer the project in a bad direction when you knew better is no different than you not knowing at all. You should speak up to help the team avoid taking a path to trouble every chance you get.

- **Always strive to be an asset**

   Be a positive contributor with all that you're involved in. The best way to do this for your team is to show up and take part. That means always making meaningful and timely contributions that help lead to the team's success. The measure of success for any team is the quality of output it generates as a whole. That output is a product of the each member's contribution combined into one result. This means that if your team fails, you fail personally for having been part of it.

When you're part of a team, you should never allow yourself the luxury of not caring about jobs that aren't directly your own. The attitude "it's not my job" has a way of diminishing everything you do. If you allow that sentiment into your work ethic, the teams you're part of will always be a bit less successful than they would have been otherwise. You may feel more in control of your workload by only completing your assigned tasks and nothing more, but in fact you are giving up control of your future. That's because when your teammates realize you only do the immediate tasks in your inbox and nothing more, they're going to send less responsibility your way. Eventually you could end up just being an expert at doing less, regardless of how much you may really know about the technology being worked on.

- **Always maintain a strong work ethic**

Whenever you demonstrate a strong work ethic, you'll often find that techs around you tend to pick up their performance too. Even if some happen to be those whose only skill seems to be avoiding work, your higher performance will usually give them impetus to improve their performance too: either they keep up or you could be their boss one day. People make all kinds of career decisions that depend more on short term needs than they do on long term goals. If you keep at least part of your focus on how your daily performance affects you in the long run, you're path will always be upward toward achieving excellence in your chosen field.

There are four good rules to follow that can help you maintain a solid work ethic on a daily basis:

Rule 1: Learn before you lean.

Work on building those skill-sets that allow you to make contributions to the team rather than relying on the effort of others. If an endpoint has been clearly defined for a job, and one should always be with any job, then you should already have an idea from the beginning what skill-sets will be required. Even if you're not the lead tech for the job, learning all you can about what's being worked on before you begin will still improve your overall contribution to the project. It may even allow you to step up to the lead role in a pinch.

The ability to learn what you need to know to get the job done is a critical skill in the IT world. Just keep in mind that project-specific learning should be measured against your long term educational goals. Most job-specific knowledge doesn't carry over to other environments or work places. While background knowledge should always be focused on the big career picture, the smaller things you learn along the way is what will make jobs assigned to you easier. Just don't burn yourself out learning more about a technology you're working on than what you'll actually need to know to get the job done. There'll always other jobs you're handling at the same time that'll require some study too and studying and learning about too many unneeded details over time will wear you out.

Rule Two: Don't let lesser opinions of others steer you down the wrong road.

Once you've gained the knowledge that gives you a solid operating base, don't let someone talk you off that base by aggressively promoting their own approach. This isn't the same as sharing ideas (whether they're good or bad). This's when you allow another tech to bully you into bad decisions. If you know their approach isn't a good one then you need to say so right away. Letting that type of situation linger only makes it worse later on. The longer you wait, the harder it'll be to avoid adding their approach to the mix, regardless of what outcome it may bring. Allowing them to believe their idea is a winning one will only make you look insincere when you finally get around to explaining the truth. It's best to be quick about responding to ideas you know won't work well…albeit in a polite and professional manner.

If you know how to do a job well, either through extensive planning or from solid experience, then you need to maintain ownership of the job no matter what counter suggestions are offered. Remember too that it's not enough to just put your foot down; you should also take the time to explain why you're putting your foot down. Get your communication skills going. Others will need to respect your differing opinion before they're willing to follow it, and nothing gains more respect than you demonstrating a solid understanding of the task at hand. If another tech's abrasiveness and swagger pushes ideas into your path and you know those ideas don't work as well as your well founded approach does, then move on past them and stay your course. Best practice over your career is to only follow the ideas you respect and to learn how to explain completely and successfully the ideas you want to promote. As you gain experience while traveling along your career path, you'll learn how to quickly recognize respectable suggestions, as well as put forth a few on your own. Just be careful not to be so confident in your planning that you become the tech bully that others around you have to deal with.

Rule Three: No one has a complete answer to everything no matter how much they know.

Don't rely so much on yourself that you become blind to that inevitable truth that plagues every confident IT professional. That is, you become so certain of your skills that you let yourself fall behind in keeping up with new knowledge. Being more certain of your skills than you should be not only seriously affects your efforts but will also make you a problem for the others you work with. If you've convinced other team members that you're the one who can handle a particular task when you're really under qualified, then you're putting the outcome of the overall job at risk. Always remain truthful about how much you can contribute to a project even if it means admitting that the input from someone else may be better than you own.

The old workplace axiom is right to suggest that, "there are no experts here, only problem solvers." Expertise can be a very addictive thing to relax a career upon. Don't be the "expert" who knows too much to accept ideas from others. All ideas are helpful in one way or another, whether they're good or bad. Book learning is important of course. It shows you how the best solutions should look after the work has been completed. However, applying that knowledge in the real world is another matter. A truth in life, and especially in tech work, is that the best responses to the toughest problems always start off as simple ideas. The more ideas a team shares, whether they're good ideas or not, the more good answers the team will have going forward. So invite ideas and share ideas. Not only will the work go faster, it'll finish smoother and be more stress free too.

Also, no matter how expert you are, if a team member was brave enough to contribute an idea to a team of skilled professionals, they deserve a professional reply even if it's only to kindly disagree.

Rule Four: Only pass the buck on rare occasions.

This means that if something doesn't work out as originally planned, don't make matters worse by expending energy on assigning blame. This's especially critical when the overall job is still underway. Blame diminishes confidence and a lack of confidence diminishes the ability to react to bad developments. Aside from being an embarrassingly useless trait to have in a competitive environment, those who are good at assigning blame are also burdening others with lousy team participation skills. It's far better to give assistance to a person who has made a mistake than to lessen their ability to contribute to the team by drumming them down.

This doesn't mean it's wrong to feel a bit angry at someone else's sloppy work or lack of initiative, especially if their issues are affecting the job as a whole. However, an excellent tech rises to this challenge by finding ways to productively communicate with the culprit instead of just alienating them as a team member. If their behavior is more of an attitude problem than a skill-set problem, you don't want to respond in kind. The problem with fighting their negative atti-

139

tude with your own is that if one bad attitude slows the job down, two bad attitudes will only slow it down further. Bad blood caused by you reacting too negatively to someone who "has it coming" will only distract from the overall effort and lead to more delays.

This advice isn't meant to be a touchy-feely, goody-two-shoes solution either. It's pure economics. There's only a limited amount of energy a team can contribute to any given job. Just as with the notion that every human being has their limits, a team of human beings has its limits too. If a team member drains these limits by their poor performance, your poor response could drain performance even further. The golden rule for being a great team member is that the project always comes before your own individual pride. After all, there'll be plenty of pride to share with everybody on the team when the job gets done well. Remember that there's a plan the team must work to keep on schedule. If a team member begins to slack off, it's the same as if you were walking on a long journey and one of your fellow travelers kept falling behind, delaying the group.

Stopping to admonish the marginal performer will only add more delays to your journey. It's far better to try to find the problem's source and then to work with them to correct it. If it's simply a lack of effort, then you must find a helpful way to increase their effort. If it's a lack of knowledge, then you should find a way to help increase their knowledge. If it's a lack of confidence, then you must find a way to buttress their confidence. Whatever it is, getting mad or frustrated isn't going to help. Your getting mad or frustrated will only be one more project distraction to be managed by the team as a whole. An overly chastened teammate will be a less productive teammate, which means the remaining team members will each haul a heavier load to get the job done.

Problem Team Members

If you have any say in how the team will be put together, it may seem a wise thing to not include problem techs in the first place. However, that isn't always possible and may also leave out a potential asset. The better choice is to include whatever cajoling and/or mentoring this person may need into your overall project timeline from the very beginning. The time spent helping them stay productive isn't wasted if it leads to regaining a potentially good teammate who'll be able to share the load eventually. It's all a matter of how you assess the risk their membership in the team brings to the job at hand.

If you have no say in your team's membership, then you may have to work with a chronic underperformer whether you like it or not. Keep in mind that you're probably not alone in your concerns. IT is such a competitive and demanding work environment that everyone eventually figures out who the problematic coworkers in the IT department are. It's the same in all team environments, regardless of the profession, and should be taken in stride as a normal part of human nature. After all, we're a tribal species and team dynamics are a natural state to us.

The best path to follow when dealing with an underperforming team member is the one that'll get the job done the most successfully. Frustration may be the first response to the person's performance, but frustration isn't going to be found along the path to success. Frustration is found on the path where you allowed the underperformer to get under your skin enough times that you've now become an underperformer too. Not allowing things to get under your skin is best practice for staying on track with any job.

It's good if the team leader has enough situational awareness to accommodate the problem person, but this isn't something you can always count on. If you're really concerned about things and the team lead is struggling with the situation, then it's OK to take matters into your own hand and confront the individual…just do it in an manner appropriate to your standing as an IT professional. After all, the success of the team is your success too. Don't forget that if you don't have a clear understanding of the root cause of the person's underperformance, you won't be accomplishing much. People underperform for various reasons so you must first *diplomatically* try to get an understanding of what their particular issues are. The best method is by speaking directly to them. Any other medium will be too trivial to have any effect. Politely asking what might be hurting their performance could yield results and hopefully allow them become unburdened by their problems too. An opportunity to vent is appreciated by anyone in a situation that leaves them unmotivated. It helps to get the gripes out of their system.

Once you're certain you have the time to get involved in the other tech's life without compromising your own work schedule, proceed with caution and talk with them directly. The optimum word here is talk. Don't e-mail and absolutely don't text. Direct verbal communication is the only effective option.

Following are common team member issues and how you can work through them to positive resolution:

- **Sometimes a team member may think the solution being implemented is not the right one**

  If they'd had more say in the solution, it wouldn't have been selected in the first place. They could be right, or they could simply be saying, "I'm tired of all this change. Why can't we just leave things the way they are for while?" In either case they aren't giving their full effort and the end result is the same. It's best to reach a tech like this by reminding them that it's a win some lose some world and that the only constant they have is the quality of their work. If they allow themselves to underperform once, it could become a habit every time they're unhappy. A reputation as a chronically unhappy tech by itself isn't much to worry about. It's the general knowledge that their chronic unhappiness causes them to be inconsistent with their work that'll kill their career. Remind them of their long term career goals. If they stick with the job, working through to the end despite their underlying opinion, then one day they may be able to have more control in their field and the reputation of an IT professional people like to work with.

- **They could be over their head in terms of technology**

  Anybody can easily become overwhelmed by the details of a complicated system. No system in the world will ever look exactly like the ones used to learn on in school. Systems are custom fit to task with real-world managerial and end-user expectations. Even standard nomenclature can become highly localized if a group of techs adapt words to their own environment. Travel from company to company as a contractor and you'll see that all the time. Even in this age of mass access to shared information, techs still develop some non-baseline computer speak that's only used in their IT department. Typically, this local nomenclature, or "shop slang," can confuse someone who just walked out of a highly structured classroom environment or just arrived from another company. It's important to be patient with the new tech. Everyone was a new tech once and got up to speed eventually. Mentoring new techs is an important part of team building. The sooner they get up to speed, the quicker they can make a contribution to the team.

Old techs can get in over their head too if they let their training lapse a bit. However, if you scoff at the legacy tech who got himself into this situation, you may risk costing yourself in the long run. Not only will you demean the confidence of someone who could serve the team with well-earned rank and leadership, but you could also have a negative personnel issue that could lead to office intrigue when you least need it.

It's far better to find a useful way to assist any IT pro fessional who needs a little help with their knowledge base for a specific job. Even senior techs would welcome this. It could be a helpful hint, a loaned certification guide, printing copies of the admin manual, providing some outright orientation or direct guidance, or even volunteering to help with a task or two. It'll be different in every case. Just remember to never leave a teammate behind and your job will always be easier in the long run.

- **Or maybe they really are just lazy**

Every profession suffers from a few lazy practitioners every now and then. It's part of the human condition. Laziness comes in many shapes and sizes and can be contagious if you let it. The thing to keep in mind when working around a lazy teammate is to not let them make you lazy too. For example, they could hold up a deliverable you need for your job. While you wait, you sit and relax and may even enjoy the break in the work routine. However, don't let the unexpected break fool you; it's always problematic.

The time a lazy teammate loses could be time you have to make up for them later. If their underachievement is only because they're a bit over their head, then help them productively. However, if they simply don't feel like working, then go light a fire under their feet. The job's endpoint won't move closer to accommodate their lack of effort, so whatever time they lose will need to be made up by someone else in order to keep the team on schedule.

Work is never avoided when a lazy person is around. It's just handed off to someone else to get done. The job will be completed regardless of who does the work. Worst case scenario is the lazy person causes such a backlog that when they finally decide to get things done, perhaps to save their job, they've already added a load of stress for their teammates to suffer through that wouldn't have been there under normal circumstances. If you know the person to be truly lazy, it'll make your life much easier if you confront them early on.

Yelling doesn't work when dealing with someone who is lazy by nature, so it's best not to bother trying it. Think about it, a lazy person has probably been yelled at their whole life for the flaw and what good has it done? Don't be too offensive either because that never helps. The lazy person will always do less by habit. They'll procrastinate until the eleventh hour and then either do a rush job or not get it done at all. Ironically, in the overall scheme of things, it may be best if they don't get the job done at all. At least that way everyone knows ahead of time the work isn't ready for the next phase. A rushed job on the other hand may look 100% complete but probably isn't. That's because rushed jobs always leave loose ends that come back to haunt you. The rushed and marginal job is what makes the lazy tech a true liability to their team.

The first thing you must always remember is to never do their work for them unless requested to do so by a boss or team leader. That usually only makes matters worse. Not only does is reinforce their negative behavior, it also adds to your workload and can cause you to be the one doing the rushed work. If rushed enough, you might actually be risking your own credibility by doing other people's work for them. The team was created for a reason. If only one guy was needed to get the job done, the team wouldn't be there in the first place. So as tempting as just giving up and getting a few of their tasks out of the way may seem to be, don't give in to the temptation. They have to do the work they've been assigned.

The second thing to remember is that nobody is actually born lazy. The concept of being born lazy is a myth. Laziness was bred out of the human genome tens of thousands of years ago. It's not a stretch to suggest that born lazy Stone Age children usually fell behind enough times to be eaten by something hungry long before they had a chance to propagate the lazy gene to offspring. Yes, this may be an over simplification of the evolutionary process, but suffice it to say that being born lazy and surviving in Paleolithic times didn't go well together. Therefore, it's reasonable to suggest that the born lazy trait was literally eaten from our gene pool. The fact is, modern laziness is a learned behavior and learned behaviors can be dealt with. If not completely un-learned, they can at least be modified enough to render the person useful as long as their team needs them to be.

Laziness is always a product of opportunity and envi-ronment. If a lazy person is surrounded by nothing but hard workers who expect them to pull their weight, they usually do. Laziness only happens with help from the outside. It's been the experience that lazy people usually have an ena-bling force nearby that's adding to their willingness to un-derachieve. This can be a coworker who's not happy with the company and constantly asking "why bother?" It could be a spouse who constantly comes to their rescue by re-minding them they're not underachievers; they're just un-derpaid and underappreciated. It can even be a worldly dis-traction that has come to mean more to them than their work does…say a new ideology they can't stop talking about. The enabling force will come in many different forms and never be the same for any two people. The point is you must identify and come between the enabling force and the lazy person in order to bring them back into the fold long enough to help the team get the job done. Don't worry, this isn't as complicated as it may sound.

You start by talking with the lazy person. Not in a con-frontational manner that just clams them up, but in a friend-ly tone that nonetheless lets them know they're part of a problem. You can't expect to talk a solid work ethic into their system, so don't bother trying. What you can do is determine one way or another who or what the enabling force is. Once you discover the enabling force, put yourself between the lazy person and their enabling force to bring them back on track. This can be done through one of two approaches. We'll call them the parallel approach and the counter approach. Both work well in the right situations.

Before you begin either approach, have a clear endpoint in mind (when their laziness will no longer be problematic). This isn't difficult so don't spend much time on it. You probably already know the endpoint because it's job related. Don't just think you want them to work harder because that'll get you nowhere. It has to be an expectation of accomplishment relative to a specific job related task or you'll only be disappointed. Remember, this's about the team and the job at hand, not an intervention to change someone's life. You only need to get the lazy person to carry their weight up to the specific point that's needed for the project to get done. Anything more than that and you'll start performing like a lazy person too; more preoccupied with curing someone else's laziness than getting your own work done on time.

Parallel Approach

The parallel approach follows the logic that by joining in with their gripes, their need for personal reassurance, an agreement with their new ideology, or whatever, you force a bond of gravity on them that can be used to pull them back in the team's direction. You work in parallel with the enabling force to exert a larger amount of influence on them and then move that influence toward your desired endpoint. This isn't a new concept. The notion of bucking someone up to get them to go along has been around since the beginning of time. The reason it's been around so long is that it works and doesn't take much time to do.

Yes, feigning interest in someone else's issues to get them to rejoin a team effort may seem like a shallow thing to do, but it works more often than not. It's also more productive than just sitting around and stewing or, worse still, doing the lazy person's work for them. The idea here is to look after your own interests by working to make sure that the jobs you're part of always get done on schedule and on budget. That's a good reputation to have and one worth a little molly-coddling of a lazy co-worker to maintain.

The sport of ice hockey has a great method of measuring a player's general trend toward helping their team. It's called the plus/minus difference. The plus/minus statistic is based on a comparison between the number of goals scored by the team versus the number of goals scored against the team while that particular player was on the ice. Sure, it's really a measurement of the team's overall performance and not just the individual player, but a player with a positive plus/minus statistic is often considered to be a positive influence on the team as a whole. The team just seems to do better whenever that player is in the game. What you want to have in your career is a positive plus/minus statistic too. That is, jobs that you're part of always seem to go better. If shoring up the occasional lazy teammate with some shallow, unctuous behavior is part of that record, then so be it. How well the project ends affects the entire team by some measureable extent and will add to your career's overall success.

Counter Approach

The counter approach also involves getting between the lazy team member and their enabling force, but instead of pulling them back toward the project as with the parallel approach, you push them back on track instead. This means making the project carry more weight and influence for them than the enabling force does. For example, if their enabler is a friend who says the company is "run by idiots, and since no one knows what they're doing, why bother doing anything at all?" then you need to convince them otherwise…at least on the "why bother" part.

Perhaps remind them that letting company dissatisfaction affect their work to such an extent will only give them the reputation of someone who can't get anything done. Often enough, a person will listen to this argument when they realize their credibility is on the line. Bosses use this approach all the time when they threaten someone's job due to a lack of performance. They put themselves between the enabling force and the lazy worker by raising the stakes higher than the enabling force can. While this can work on occasion, it's not something a co-worker has the authority to do so; you'll probably need to find something similar to the boss's mandate, but less threatening. Use your best judgment too since bad blood within a team could add more stress to a project than the one team member's underachievement was.

If neither of these two approaches are successful and the lazy teammate continues avoiding their fair share of the work, then the last ditch approach is to voice your concerns to the team leader or to upper management. This may make you feel a bit guilty, but always remember plus/minus difference and its impact on your career. Even this last ditch step can go a long way toward helping you develop a reputation as someone who makes jobs go better whenever you're around.

- **There may be a time when an outside influence is simply too strong**

A team member can suffer an outside influence that's difficult to overcome even with your help. That's when something major has happened to them outside the workplace. Sometimes even excellent IT professionals can perform in a lazy manner if something disruptive comes into their life and distracts them from their job. It would be nice to suggest only positive events cause this kind of distraction, for example, a recent marriage, a new baby, a child who's achieving great success, winning the lottery, and so on. However, negative life events can cause them too, such as divorce, the death of someone close, a child in trouble with the law, financial problems, psychological disorders like depression, or a physically disabling event like a car accident. These can knock even the best IT professionals off their game. A teammate in this situation isn't really lazy; they just can't get the work they're assigned done on time.

The cruel truth is it makes no difference how serious or painful the distraction may be. All life distractions will have the same negative impact on your team's workload. As such, you have no pressing responsibility to deal with the teammate's distractions other than to find a way to make up the lost ground caused by those distractions. In fact, unless you're a close friend of the teammate outside of the workplace, it's probably best if you don't get too involved. Efforts to help a person deal with a personal crisis just to make your job go easier will at best be insincere. This's one of the few times when adjusting workloads to accommodate someone's underachievement is advisable. In fact, a savvy team leader may have already built redundancies for this kind of issue into the job plan.

Time is finite. The more time you give to your friend's issues the less you'll have for your own needs. Sacrificing for a friend may be heroic, but if you're a member of a team, you have to be heroic to your teammates too. Becoming overly preoccupied with your friend's problems could mean your team now has two troubled members to make up for.

Balance is critical here. If there's no immediate personal commitment, then it's best to focus entirely on the job at hand. If there's a need that you feel must be accounted for, don't forget that your team members are counting on you same as your friend may be. The excellent IT tech will always stay on schedule. If you want to spend more time helping your friend, then balance it with more time on the job. This may even require adding extra time to your work day to get both done. Again, it's your choice and there's only twenty-four hours available in each day. Staying on task with the job should be your first priority always. It's your professional calling and what you do for a living. What's more, those lifeless, emotionless computer systems aren't going to wait for you or your troubled friend to regain focus. They'll just keep rolling along oblivious to life's conditions.

Finally…

Being a good team member means contributing your part in any working situation you're sharing with others. Doing your job completely and on time is a measure of how good a tech you are individually. This should always be your focus and is a large part of your professional bearing. When being a part of a team, don't compromise your own standards. Trust your own judgment first. This means maintaining a situational awareness of everything going on around you. If you have experience and are feeling uncomfortable with the way a job is moving along, you can always fall back to that experience. More importantly, convey your experience to others in your team so the whole group can benefit. If a boat in which everyone holds an oar accidentally floats over a waterfall, it's everyone's fault regardless of who was trying to row at the end. You were made part of a team for a reason. Don't undersell yourself. Contributing is contributing. It's not just getting assigned tasks done on time. It's delivering an input that determines how well those tasks get done on a professional level.